MANUFACTURING INFORMATION SYSTEMS

Implementation Planning

MANUFACTURING INFORMATION SYSTEMS

Implementation Planning

Robert A. Gessner

A Wiley-Interscience Publication

John Wiley & Sons

New York Chichester Brisbane Toronto Singapore

Library of Congress Cataloging in Publication Data

Gessner, Robert A.
Manufacturing information systems.

"A Wiley-Interscience publication."
Bibliography: p.
Includes index.
1. Production planning. 2. Management information systems. I. Title.

TS176.G496 1984 658.5 84-7408

ISBN 0-471-80843-1

Printed in the United States of America

10 9 8 7 6 5 4 3 2 1

PREFACE

Most progressive manufacturing companies have a long-range plan defining where the company wants to be in five to ten years. Unfortunately, it seldom defines what has to be done tomorrow, as the first step in achieving the long-range plan.

This book defines the methodology for building the plan to define what has to be done tomorrow.

It describes how you, as a session leader, will lead a select group of manufacturing managers through the methodology for four to six weeks. You will produce an implementation plan that is very detailed up front and is more general as it approaches ten years in the future.

Your implementation plan will be based on a detailed problem analysis and a general system design. It will be load leveled so that it is not a "wish list" but can be realistically installed. It will be cost justified.

In addition, you will receive guidelines on how to prepare for a study, what material to use in presentations, how to structure the study report, and much more.

This basic planning methodology has been successfully employed in hundreds of companies. The approach works.

ROBERT A. GESSNER

Kennesaw, Georgia
May 1984

CONTENTS

PART ONE. FIRST STEPS IN PLANNING A SYSTEM STUDY

1. Introduction **3**
Who Can Benefit from This Book, and How 3
Overview of the Book's Content 4
Why This Approach Works 5
Some Keys to Success 6

2. How to Sell an Implementation Study **9**
Defining Your Audience 9
Request for Data 10
Setting Up the Sell Presentation 11
Importance of Recruiting an Executive Sponsor 12

3. Preparing the Initial Presentation **15**
Purpose of the Presentation 15
The Presentation Audience—Who Should It Be? 16
Presentation Media Considerations 17
Examples of Presentation Content 18

4. How to Make Effective Presentations **39**
Some Do's and Don'ts 39
Avoiding Prejudged Conclusions 41
Obtaining Commitments 42

PART TWO. HOW TO LAUNCH THE STUDY

5. Groundwork: Laying Strong Foundations **47**
Guidelines for Measuring Progress: Evaluating Results of a Study Day 47

How and Why to Log Team Attendance 48
Checkpoints—The Interim Reviews 49
Preplan the Final Report—And Save Time Later 50

6. Start-up Activities **55**
Review of the Executive Sponsor's Presentation—
Why the Review Is Needed 55
Doing a Company Analysis Questionnaire 56
Defining Functional Relationships 57

PART THREE. CONDUCTING THE STUDY: THE FOUR PHASES

7. Phase I: Problem Definition **63**
Purpose of Problem Definition 63
Brainstorming for Interview Issues and Personnel 64
How to Set Up and Conduct Interviews 64
How to Group Detail Problems by Company
Functional Areas 65
How to Restate Groups of Detail Problems as Major
Problems 66
Prioritizing Major Problems 66
Defining the Study Team's Objectives in Response
to Major Problems 67

8. Phase II: General Application Design **73**
Purpose/Objective of This Phase 73
Defining the Current Operational Techniques 75
Defining the Desired Operational Techniques
to Solve Key Problems 76
The Definition of Goals to Achieve the Desired
Operational Techniques 79
Establishing the Criteria for the Selection of Tasks 80

9. Phase III: Implementation Planning **103**
Purpose/Objectives of Implementation Planning 103
How to Subdivide Goals into Tasks 104
Building a Task/Goal Dependency Relationship Chart 106
How to Assign Tasks to Individuals and Estimate Task
Durations 106
How to Load Level Tasks by Person Within Departments 108
How to Formalize the Implementation Plan 111

10. Phase IV: Justifying the System **129**
Why Justification Is Essential and Ways in Which Lack of It Can Lead to System Failure 129
Reviewing Potential Benefit Areas to Evaluate 130
How to Document and Summarize Tangible Benefits 131
How to Define and Validate Intangible Benefits, and Why They Are Important 133
Reviewing Potential Cost Areas to Evaluate 133
How to Document Cost Requirements 135
Preparing a Cost-Benefit Summary 135

PART FOUR. FINAL STEPS: THE BIRTH OF A SYSTEM

11. Preparing for Effective Project Management **157**
Objectives for Project Management 157
How to Develop a Project Manager's Job Description 157
How to Set Up Procedures for Task Tracking and Projection 159
How to Set Up Procedures for Tracking of Actual Costs and Benefits, and Projection of Further Costs and Benefits 161

12. Preparing the Final Report and Final Presentation **179**
Preparing the Final Report 179
Preparing the Final Executive Presentation 180
Conducting the Final Executive Presentation 182

APPENDIXES
A. Sample Nomination Form 185
B. Sample Company Questionnaire 197

PART ONE

FIRST STEPS IN PLANNING A SYSTEM STUDY

CHAPTER ONE

INTRODUCTION

WHO CAN BENEFIT FROM THIS BOOK, AND HOW

The amount of benefit that you can gain from this book obviously depends on who you are and why you are reading it.

If you are a college student, you will obtain knowledge about methodology that defines, designs, plans, and justifies systems to "fix" problems related to how a business operates. Your real benefit will occur when you actually put the methodology into practice. This process will provide more exposure to business operational practices in a few short weeks than you could obtain by working for a company for years.

If you work for the company to which you would like to apply this methodology, your benefits are the following:

1. As an employee, you have a position in a specific department, which let's say is Production Control. Do you know how the Design Engineering Department functions? How about controlling orders on the shop floor? At the conclusion of the application of this methodology, you will have a fairly good understanding of the total company and this makes you a more valuable employee.
2. In the past, you may have been a name and badge number among hundreds. In the process of doing a study like this, you will rapidly find that top executive attention is focused on you. Suddenly the Vice President of Manufacturing knows who you are and what you can do. Note that this benefit applies to your whole study team as well as you. I conducted one study on which I had six team members. Within 30 days of the end of the study, five of those six team members were promoted. (I was not an employee of the company.)

You might be doing consulting work for a company. If you work for yourself, your potential benefit is obvious, since the going rate for this type of work is $800 to $1200 per day, plus expenses. Since the average study takes six to eight weeks, you can add up the numbers.

Probably, the key benefit, regardless of your position, is that this book will allow you to display a professional approach on your first study. I have done (and helped others with) "first" studies. The use of the word "floundering" in describing the techniques used by some session leaders on first studies is a gross understatement. A lot of trial and error went into "How to Do It Right" and, as a result, this book.

OVERVIEW OF THE BOOK'S CONTENT

This book does not address the question "Should we do something to fix our problems or just do nothing?" The assumption is that the company with which you are working has recognized that one or more problems exist and wants a "fix" to the problem(s).

The structure of the book is the same as the process that you will use to conduct a study. A study means that you and four to six key company personnel will work six to eight hours a day, four to five days a week, for six to eight weeks, doing the following:

Defining the problems in the company
Developing a general application design
Building an implementation plan for the design
Cost justifying the total plan

The process is very logical. Before you can do a reasonable job of cost/benefit analysis, you have to have an implementation plan defined so that you know what is going to occur and when. The implementation plan obviously cannot be defined until you have completed the application design and know what it is that you want to implement.

The next step is a tricky one. Do you have to do problem definition before you do application design? Most companies will tell you, "Of course not, we know exactly what the problems are." And they are usually wrong. One company insisted that their key problem was capacity requirements planning. We could have addressed that directly, but we didn't. We went through the problem definition phase. What we found was the following:

1. Capacity requirements planning (CRP) was a problem.
2. Material requirements planning (MRP), which feeds CRP, was inconsistent and inaccurate in its output.

3. Three different product structures existed for any one product—one for costing, one for engineering, and one for manufacturing.
4. Inventory accuracy was about 50%.

Since product structures and inventory are key drivers for a good MRP system, the final plan called for fixing those things first. Then we cleaned up remaining MRP problems. Finally, we addressed CRP problems, but in the implementation plan the CRP solutions did not occur until four years into the plan. We would have been dead wrong to build a plan to fix CRP problems now, when all of the systems that fed CRP were defective.

In addition to the core of this book—doing the study—other related topics are covered. Such as:

How to sell the idea of a study
Preparing your initial presentation
How to make effective presentations
Laying the groundwork
Start-up activities
(The study)
Preparing for effective project management
Preparing the final report and final presentation

The book presents a very structured approach, which takes you from selling the idea to implementing the results.

WHY THIS APPROACH WORKS

First let's consider why some other approaches do not work.

Method 1. Develop long-range company goals. Assign major responsibilities to department managers. Measure progress in monthly staff meetings.

Results. Each department develops its own plan. No plan integration between departments. Department problems are prioritized, not key company problems. Department empires are protected. The program is "laid aside" when the next hot issue comes up because this approach is seldom cost justified.

Method 2. Hire a consultant to come in and evaluate the operation.

Results. You may get a recommendation such as "lock up your inventory and improve your inventory accuracy." This means nothing, since you probably already understood the solution in general terms, and you still don't have a plan on how to do it. (You may also get a recommendation to do this type of study, which is obviously okay.)

Method 3. Assemble a task force to "fix the problems." This is usually a combination of employees and managers who have neither an approach to obtaining the solution nor the time to dedicate to the effort.

Results: Some general recommendations, no implementation plan.

There are several good reasons why the methodology presented in this book works where others fall short.

You do not "assume" what the company problems are. You establish a clear identification and prioritization of key problems.

You constantly review progress and direction with top management throughout the study effort.

You build an implementation plan which:

Addresses the solutions to key problems

Is front-end detailed down to who will do what and when

Is a total effort integration.

You cost justify the effort. If this plan saves *X* dollars, it can be properly evaluated against any other hot project that saves *Y* dollars. Projects that are not cost justified are dropped. Something else can always be made to "appear" more important if cost justification does not exist.

History is probably the best proof of the success of this approach. Most companies (80 to 85%) that received our advice implemented the plans that my study teams developed. The approach works.

SOME KEYS TO SUCCESS

Some very important keys to success exist. Should one or more of these keys be lacking, either while you are initiating your study or while you are conducting it, be on the alert for potential problems.

Top management must be committed and involved. If you do not have top management support, walk away from the study because this is guaranteed failure. You may have a production control manager or a data processing manager who thinks that building a manufacturing information system is the greatest idea in the world. If you do not have the support of the president and the vice president of manufacturing, you have an exercise in futility.

Active participation by the management of every involved plant is required. You do not want middle management giving the VP hundreds of reasons why "your" plan won't work. They have to be involved with the development of the plan. Then they will be telling the VP why they need a resource to make "their" plan work.

Manufacturing personnel must be available during the study and when necessary during the implementation. During the problem definition phase you will conduct interviews with manufacturing personnel at all levels. If you set up a meeting and get a reaction like "I can't make it, I'm too busy," you have several problems. You do not have full top and middle management support, so your study will lack the proper inputs required. Your implementation plan will not run smoothly because when it comes time for that same person to perform a specific task, he or she will be "too busy."

A full-time project manager is required. In most companies, this is a new position—another salary. This requirement should be made clear up front. If you get evasive reactions to this requirement, like "Yeah, well maybe we'll do that," you are in trouble. A company that will not support this expense will probably not spend the money to train its personnel on new material handling techniques, for example.

A training coordinator is required. This may be a full- or part-time position. In a small company, it may be the project manager or someone in personnel. In a large company, it is a full-time position and probably already exists.

Data processing personnel must be made available when specified. All too often, when DP is requested to install a new function, their reaction is to say, "Sure we can do that—in two years." Why are they so busy? In the average company, they are doing two things.

Adding improvements (functions) to existing applications, such as accounts payable

Maintaining existing applications, such as removing defects and improving performance

Your plan and DP's current work load have to be put in perspective. For example, how much benefit will the company get from a new terminal display screen in accounts payable versus a software module that will allow a 20% reduction in inventory?

Usually, additional data processing equipment will be required to support your plan. For some unknown reason, the acquisition of a new forklift in the warehouse is a different type of capital expenditure than the acquisition of a new disk drive in the DP department. Make it clear from the beginning that your plan could very likely identify both the forklift and the disk drive as being necessary and cost justified, otherwise they wouldn't be included in the plan.

Your final key to success is a commitment to continuous manufacturing systems education. The major reason, for example, that MRP systems fail, is a lack of understanding of what the system is doing—a lack of education. This goes for all personnel from the president down to the milling machine operator, obviously concerning different types of education. If the education is not built into your plan (or not completed as scheduled), the personnel implementing the plan will not know why they are supposed to be doing something and therefore will not do it—and your implementation stops.

Of the studies that I have done, two have failed because a key to success "went away."

Failure 1. The plan was good. All management was fully behind it. The implementation was going well. The projected benefits were being realized. The project manager was doing an excellent job. Management promoted the project manager. The implementation stopped. Four months went by. The company hired (from the outside) a project manager. The implementation never restarted. The plan was not the new project manager's plan. He spent the first year just learning the company operations.

Failure 2. The plan was designed in mid-1981. It was fully supported and the implementation was going well—until the recession hit in 1982. Expenses had to be cut back. An arbitrary decision was made to eliminate all travel. This also eliminated all travel to classes and conferences, which had been designated as educational requirements that were vital to the success of the implementation. The VP of Finance convinced the VP of Manufacturing that the education could be carried out through books and local seminars. By June of 1982 the implementation had stopped. Now (hindsight being fantastically accurate) the company management recognizes that a year has been lost, and they want to rebuild the plan.

From the time that you first talk to the top management of the company until you finish the study, you should monitor your progress to be sure that you have the "keys to success." If they are not there initially, or vanish during your study, stop and walk away—obviously after explaining your reasons to top management.

As a private consultant, you might disagree. You might say, "What the heck, for $1000 per day I'll finish the study. I don't care if it's implemented or not." Okay for that study, but how many failures can you use as references for future prospects?

CHAPTER TWO

HOW TO SELL AN IMPLEMENTATION STUDY

DEFINING YOUR AUDIENCE

Who you are going to sell a study to obviously depends on who you are. If you work for the company under consideration, then you can skip to "The Importance of Recruiting an Executive Sponsor." If you work for a company that has a marketing staff in the field, they are defining your audience for you. In this case, skip to "Request for Data." If you are an independent private consultant, you need to do some preparatory work.

As a private consultant, you do not want to spend your time knocking on doors. You should first identify one or more specific industry subsets in which you are knowledgeable. Next, identify the companies that fall within that subset that are within the geographical territory that you want to cover. Now divide these companies into categories of:

Group 1. Growing by leaps and bounds all by themselves (and have the expertise to do in-house what you are selling).

Group 2. Not growing (or losing ground), and "family" owned.

Group 3. Not growing (or losing ground), not "family" owned, and the general economic situation has no bearing on the lack of growth.

Group 4. Not growing (or losing ground), not "family" owned, and the general economic situation has a direct bearing on the lack of growth.

What you really want is a high success rate. In Group 1, you will seldom have a chance. If they can do it themselves, they do not need you. This leaves you with the other three groups, which you may consider to be "trailing edge" companies. There is nothing wrong with that. They are the folks that need the most help and will be the most receptive to your offer.

Group 2 is one to be careful of. They may have lots of problems and need lots of help, but you may be faced with the syndrome of "I did it this way 20 years ago, it worked, and I'm not going to change now." You have to be positive that your executive sponsor for the study is the person that can enforce, not only your study, but your implementation plan. Without that, you may do the study, but the "family" may decide not to use it. Okay, you got paid for the study, but do you want a loser as a reference for your next study?

Group 3 holds the highest potential for success. These folks have problems, normally know that they have problems, and would like to have their problems fixed. When they understand what you can do with a study, they are an easy sell. They are faced with "Fix the problems or go out of business." To fix the problems they will need your study's implementation plan.

Group 4 is a mixed bag. I have called on accounts that had everyone laid off except top management, due to the 1982 recession, and they wanted to do studies. The thinking process was "If things pick up, we're ready. If they don't, we've spent a few dollars, but we're out of business anyway. So they do studies. On the other side is the company who says "I can't do a study now. All my people are laid off, and the key people that you would need for a study are busy running what's left of my operation." You don't want this company for a study. This person's foresight extends all the way out to next week. There is a low probability of implementing your plan (if you did do a study), because "things would change" as the economy picks up.

REQUEST FOR DATA

You identified a list of potential accounts and you would like to narrow it down to the highest "potential hit ratio" for successful studies. (I measure a successful study as one that is implemented.)

You need more data to make your evaluation. Obtaining an annual report or asking for a plant tour are things you should always consider. In addition, you should have your potential customer complete the Nomination Form and the Company Questionnaire (examples in appendix). These will provide you with the data to assess:

1. What type of company is this? Does it employ a make-to-stock or an assemble-to-order production process?
2. What types of controls/systems does the company have in place today?
3. Where does the company think that they are going in terms of manufacturing systems growth?

4. Is this company leaning toward flexible, flow, repetitive, or some other manufacturing philosophy? Will some systems currently in place not be needed in the future?
5. Are there some basic management policies that are obstructing the development of improved control systems? (Be careful of this one.)

You are going to find out, when you ask your customer to complete the nomination form, that it requires basic data which anyone can provide—even a marketing representative, if you work for a company that has a field force. However, when you try to get the company questionnaire completed, there is seldom one person in the company who can complete it. Seldom can a field marketing representative complete it. No one person in the account can do it, because it covers a multitude of functional areas. Give it to your executive sponsor and have him or her rotate it through the personnel that are qualified to obtain and provide the information. Note that the average company receives a very real benefit from completing the company questionnaire—they find out how their company runs.

SETTING UP THE SELL PRESENTATION

When you say to a potential customer (or your manager, if you work for the company) that a study should be done, the first question is "Why?" Why me? Why should I expend the resource? (The salaries of the study team are a real expense). What will I get out of it?

Well, you obviously cannot say that the customer will have a more effective way of doing material requirements planning by June 16 of 19XX. At this point you are not even sure of his problems, much less the solution to those problems and when that solution will occur.

So what do you need? You are going to do a pitch. You are going to try to sell a person (or group) on buying your ideas. You need two things.

Overviews and examples of what you are going to do throughout the study. (See Chapter 3—"Preparing the Initial Presentation.")

References. When did who do what? What implemented? Can they be called? What were the net benefits?

If you are doing your first study, the references are tough to get. You could lie, but I wouldn't recommend it. It is too easy to check out if someone wants to. Play it straight. On your very first study, admit that this is a process that is going to be mutually beneficial. You'll get a reference and they (the customer) will get a solution to their key problems. Charge less or charge nothing (if you are a consultant). It will be worth it in the long run. Build your references as you progress.

IMPORTANCE OF RECRUITING AN EXECUTIVE SPONSOR

This is a very important point. If you consistently select the wrong person as your executive sponsor, you will have few study successes.

The selection approach will vary by your level of comfort in regard to executive management and who the company tells you has the decision-making capability.

Prior to the start of the study, the selected executive sponsor should:

Understand and approve of the approach that you are going to take in the study

Be able to commit the required study resources

Be able to commit the resources necessary to make the implementation plan a reality

Be aware and approve of the participation that will be expected of him or her

Prior to your initial presentation, you should have an idea from the nomination form who the executive sponsor might be. If you think that the wrong person has been chosen, do not try to have a change made until the conclusion of your initial presentation. Do, however, try to have your choice of executive sponsor in attendence at the presentation. More often than not, the company appreciates your suggestion and allows a change.

The correct choice for an executive sponsor is going to vary by company size.

1. In a small company (200 to 500 employees), the executive sponsor should be the president.
2. In a medium-sized company (500 to 5000 employees), the VP of Manufacturing is a good choice.
3. In a very large company, your choice will vary according to the scope of the study. If the study actually addresses the entire company, then the sponsor is the person who can enforce the implementation throughout the entire company. However, if the study addresses one plant out of many, then it is the person who can enforce the implementation at that plant, normally the plant manager, who may also be a vice president.
4. Seldom should you select anyone associated with data processing as your sponsor. Most DP departments are viewed as service organizations that support the departments that really make the product (and the money). The average DP department can suggest improvements to a manufacturing operation, but seldom can they enforce or control an implementation.
5. Do not select a "new hire," regardless of position. That person is still learning the workings of the company and is not yet politically in control.

There will be times when you will be tempted to go ahead with a study even though you know that you have the wrong sponsor. Do not do it. You will be better off to walk away from it and start again at another company, than to waste weeks and gain little more than frustration.

CHAPTER THREE

PREPARING THE INITIAL PRESENTATION

PURPOSE OF THE PRESENTATION

This is a selling function. Whether you are trying to sell a company on a study so that you can establish a consulting contract, or you just want to instill improvements in the company for which you work, the process remains a selling function.

There are only two points that you have to make in this presentation:

1. The company needs the study.
2. You are the most qualified person to conduct the study.

To convince your audience (the company) that they need the study, you have to discuss what they will get from doing the study. This is normally a brief review of what will be prepared as a final report, including examples of report subsections, such as "The Management Summary," "The Implementation Plan," and "The Project Management Plan."

From the nomination form and the company questionnaire, you should have obtained some indications as to potential problem areas. Raise these as questions. For example, you might say, "When I asked about your inventory accuracy, I was told that it is probably around 50 to 75%. Now, we both realize that the percent of accuracy is low, but what really concerns me is the lack of knowledge about the exact amount. It is not 63%. It is 50 to 75%. Is this a concern of yours, and if so, is this the type of concern that we might address in the study? Can anyone venture a guess as to what it might mean to you if your study team could put a plan in place to raise your inventory accuracy to 95%, say in two years?"

A statement like that does several things:

1. It identifies a problem with which they can associate.
2. It demonstrates your awareness of their problems.
3. It suggests that you have a potential solution.
4. It presents the solution as being theirs, not yours. Remember you said, "*Your* study team could put a plan in place."

You've done your homework. You understand the company. You display your understanding during the presentation, and the company needs the study.

Now, why are you the most qualified person to do the study? You must convey these attributes:

1. You are the only one who understands the study methodology.
2. You understand the industry.
3. You have demonstrated that you have an understanding of this company's situation.
4. You have been successful on previous studies.

To establish the last attribute, you may be asked for references. If you have them, use them, but first obtain approval from the companies that you are going to submit as references. If you don't have references, admit it if asked directly and negotiate your fee accordingly.

THE PRESENTATION AUDIENCE—WHO SHOULD IT BE?

If possible, you should be involved in the selection of the audience for your presentation.

The audience should obviously include your executive sponsor. It should also include your potential study team, if possible. In addition, it should include all other management levels that might be affected by the study.

Let's say that you find yourself in a situation where John Smith, VP of Manufacturing, is your executive sponsor. You want John at the meeting. You also know that Bill Brown, the Controller, is dead set against spending money for a study. Do you want Bill at the meeting? Darn right you do. Do not ever try to bypass politics by talking to only the "right" people. If there is a potential that Bill can kill your study, or cause enough disruption that the results will be worthless, you want that out in the open now. Get both John and Bill in the meeting. At the end of the meeting, you will ask for approval to proceed. If you get Bill's approval (or do not get a disapproval), you can proceed. If you get disapproval now from Bill, then you are entitled to walk away from this project until the company gets internal agreement on what they want to do.

PRESENTATION MEDIA CONSIDERATIONS

The medium that you select has to be one that you can use comfortably. Your basic choices are as follows:

None. Just stand and give your presentation. This is the least expensive, since it involves no development of materials or handouts. However, since there are no materials, you have nothing to leave for your audience to review after the meeting, and this is a real disadvantage. Also, without illustrations of some sort, you often must attempt to verbally describe something like a break-even analysis chart to someone who has never seen one.

Slides. This is a very professional presentation. It also allows you to insert actual photographs of referenced companies, previous study facilities, and other material. The cost of this approach is somewhat higher than most others, and the slide presentation also tends to be less flexible than some other methods.

Foils. This is the medium that I have always used. The preparation cost is low. I would have 500 foils and from that basic set, tailor a presentation easily and quickly for any company. The foil medium also allows you to use examples from final study reports of previous studies. I construct a set of foils for every step of an actual study with multiple examples (from previous studies) for each step. I then tailor an initial presentation easily and quickly. (Sometimes foils are referred to as overhead transparencies.)

Flip Charts. Prepared flip charts are a bother to carry and are difficult to easily "tune" to the needs of a specific company. Professionally prepared flip charts are also somewhat expensive. Another approach with flip charts is to start with blank paper and build the pitch as you give it. This approach is very professional and illustrates that you really know what you are talking about. However, you'd better really know what you are talking about before you try it. The additional slight disadvantage on flip charts is that, if you leave the flips that you just developed with the account, the customer has to have the flips typed and distributed. That time lag may act as a cooling off period for the enthusiasm to get a study started.

Video Tape. This is probably the most expensive medium and the most worthless for this purpose. The company people are interested in you. They want to interact with you during the presentation. They can't interact with a video tape, even if you are there and you are on the tape. It is impersonal and cold. It is a good approach, however, for a marketing staff that is trying to sell your services.

If you are just starting, you should try different types of presentation medium and settle on the one with which you are the most comfortable. Remember that you are trying to appear relaxed and in total control. You will not be able to do that if you are using someone else's material or a presentation medium with which you are not comfortable.

EXAMPLES OF PRESENTATION CONTENT

When you give your presentation, there are several key items that you should cover:

Executive commitment to the total program
Qualified team members on the study
A project manager designated prior to the study start
An agreed-upon work schedule
The availability of resources for the study, including facilities and supplies
The availability of pre-study activities, such as a plant tour, study team education, and documentation.

In addition, you want to illustrate what you are going to do during the study and what the company will get from doing the study.

I would frame my foil presentation as follows. (Note: Copies of your foils should be prepared as handouts.)

FIGURES*

Figure 3.1 is a sample cover page. You should tailor the sample by replacing the "John Doe Company" with the actual company name and also putting the date of the presentation into "Today's Date."

This is the general outline of the presentation. You will be talking about the study phases, a sample agenda, a sample final report, and, finally, the study requirements.

How to Plan the Implementation of Manufacturing Information Systems for The John Doe Company

1. The Study Phases
2. A Sample Study Agenda
3. A Sample Final Report
4. Study Requirements

Your Name
Your Address
Your Phone Number
Today's Date

FIGURE 3.1 Presentation cover page.

Figure 3.2 illustrates the points that you are going to cover under "The Study Phases." As you have this foil on the overhead projector, you should summarize the content of each of the four steps. Do not go into great depth. A minute or two on each point at this time is adequate.

The Study Phases
Problem Definition
General Application Design
Implementation Planning
Justification

FIGURE 3.2 The study report.

*These and all subsequent figures are located at the end of their respective chapters.

Figure 3.3 is the sample study agenda. Now you can spend some time on each of the steps within each study phase. Do not, as you read this, be concerned that you may not understand the meaning of each step. When you have completed this book, you'll have that understanding.

You are now going to review the anticipated product of the study—the study report. While this report may be viewed as "only a book," it is actually the documented plan of how to change the business for a more profitable operation. We discuss the foils on the study report one section at a time.

Sample Study Agenda

Problem Definition

1. Review the Total Study Methodology.
2. Review the Company Questionnaire.
3. Interview Personnel to Define Detail Problems.
4. Categorize the Detail Problems by Company Functional Areas.
5. Define Major Problems.
6. Prioritize Key Problems.
7. Define the Study Team's Objective.
8. Define Current Operational Techniques.
9. Define Desired Operational Techniques.
10. Define the Goals Necessary to Achieve the Desired Operational Techniques.
11. Sequence the Goals by Dependency Relationships.
12. Estimate Goal Durations.
13. Calculate Goal Start–Stop Dates.
14. Identify the Time Fence for Task Selection.
15. Conduct an Interim Review.

Implementation Planning

16. Subdivide Goals into Tasks.
17. Define Task and Goal Dependency Relationships.
18. Define Task Responsibilities and Estimate Task Durations.
19. Load-Level Task Labor Requirements to Provide a Realistic Implementation Plan.
20. Extract Task Dates from the Load-Leveling Bar Chart.
21. Prepare an Application Summary Chart Which Reflects When Major Systems (as MRP or Forecasting) Will Be Implemented.
22. Document All Developed Implementation Planning Data.

Justification

23. Review of Potential Benefit Areas.
24. Define Potential Intangible Benefits.
25. Define Potential Tangible Benefits.
26. Define Intangible Benefits.

FIGURE 3.3 Sample study agenda.

27. Time Phase Tangible Benefits by Quarter.
28. Review Potential Cost Areas.
29. Define Cost Requirements.
30. Time Phase Each Cost Requirement by Quarter.
31. Summarize All Costs by Quarter.
32. Prepare a Break-even Calculation.
33. Make a Break-even Analysis Chart.

Wrap Up

34. Define Project Management Procedures.
35. Finalize the Study Report.
36. Conduct the Final Review.

FIGURE 3.3 *(continued)*

Figure 3.4 is the study report cover page. It includes the company name, study dates, and study participants.

Manufacturing Information Systems
for
ABC Company
August 19XX

Session Scope

Month, Day, Year to Month, Day, Year
Four Days per Week

Session Participants

Your Name—Session Leader
Name—Master Scheduler
Name—Manager of Materials
Name—Product Development Engineering Manager
Name—DP Director
Name—Plant Manager
Name—Production Control Manager

FIGURE 3.4 Study report cover page.

Figure 3.5 is the study report preface. It is a summarized table of contents, which very briefly tells the reader about the total structure of the report.

Report Preface

The ABC Company Study Report has been structured into four independent sections:

I. Management Summary
II. Implementation Plan
III. Project Management
IV. Appendix

The *Management Summary* provides an overview of what was done, what was produced, and a summary of the Cost Benefit Analysis.

The *Implementation Plan* is the final session product. This section is the working implementation plan. It identifies tasks to be performed, personnel, responsibilities, durations, and start–stop dates.

The *Project Management* section provides the input to project management to evaluate actual performance against planned activity.

The *Appendix* defines what was done during the session and what products were produced at each step of plan development. This section is included for two reasons:

1. To indicate to the reader, the methodology employed to arrive at the tasks included in the implementation plan
2. To act as a guide for future replanning efforts

Note that although the four sections are independent, to minimize redundancy the data in Sections I, II, and III have not been repeated in Section IV at the point where they were derived.

FIGURE 3.5 Study report preface.

Figure 3.6 is the study report management summary. As you look through Figure 3.6, you may think that it is a short management summary. You're right. You should hold the total management summary to a maximum of five or six pages. The intent is to explain what was done and what potential benefits exist as a result. Note that the last two pages of the management summary are overview charts concerning the general implementation and break-even analysis.

Management Summary

In April, 19XX, the ABC Company decided to participate in a joint study activity with (your name) called a manufacturing information systems study. The purpose of the study was to develop a detailed plan to build our manufacturing strengths.

The study began on June XX and concluded on August XX—an intensive six-week period. Six key people from division and group participated in the study on a full-time basis. (Your name)'s role was to support the company through the use of a proven, successful methodology for producing an implementation plan. The study addressed the manufacturing needs of all plants.

The functional areas addressed during the study were:

Business planning
Forecasting
Demand analysis
Production planning
Master schedule planning
Order entry and inquiry
Inventory accounting and control
Bill of material structuring
Material requirements planning
Shop order release
Shop floor feedback
Routings
Facilities
Purchasing
Receiving
Product costing
Plant maintenance

The tasks undertaken were as follows:

Interviews with personnel from all plant locations to identify problems
Grouping of problems by functional area and key problem identification
Constructing a general application design
Establishment of goals to be implemented
Development of a logical sequence of implementing the goals considering dependencies and duration

---

FIGURE 3.6 Study report management summary. *(Continued on next page)*

Definition of detailed tasks required for goal implementation
Development of a detailed implementation plan with task responsibility assigned by department
Estimation of the benefits and costs to implement the goals, including a break-even analysis
Preparation of the final report and executive presentation

The proposed systems are on-line to the central processor and are supported by an integrated data base. Key elements are:

On-line data update and retrieval
Editing and error detection at the transaction source
Automatic generation and transmission of action messages
On-site, user programming capability

The key elements of concern were:

Management of inventory
Feedback from the shop floor on the status of orders, material, and labor
Management of costs

Tangible benefits proved to be difficult to define due to a lack of data as to our status today. This undoubtedly is the single largest justification for the implementation.

Cumulative net benefits accrued from the new systems will approach three million dollars by the fourth quarter of 19XX. The break-even point is January 19XX. Project slippage will delay net benefits of $250,000 per month.

The following are prerequisites to the success of the plan:

Top management commitment and involvement
Major participation by every involved plant and its management
Assignment of a full-time training coordinator
Availability of specified manufacturing personnel
Availability of specified data processing hardware
A full-time project manager
Availability of data processing personnel
A continuous commitment to manufacturing systems education

FIGURE 3.6 *(continued)*

IMPLEMENTATION PLAN OVERVIEW

Key Areas	Preparation Starts		Preparation Is Finished		1981		1982			
	Task Number	Date	Task Number	Date	3	4	1	2	3	4
1. Establish steel service center	99.115	6/8/81	99.117	7/22/81						
2. New Engineering Procedures	99.120	6/8/81	99.24	4/27/82				Refinement		S
3. Status reporting and product costing	99.142	6/8/81	99.160	6/28/82						S
4. General procedures	22.2	6/17/81	22.11	11/24/81						S
5. Material planning	99.121	6/17/81	99.31	7/26/82						S
6. Material control	99.95	6/21/81	99.83	2/2/82						S
7. Ministudies	4.9	10/5/81	4.2	12/22/81						
8. Replan MIS study	99.141	2/2/82	99.141	3/2/82						

FIGURE 3.6 *(continued)*

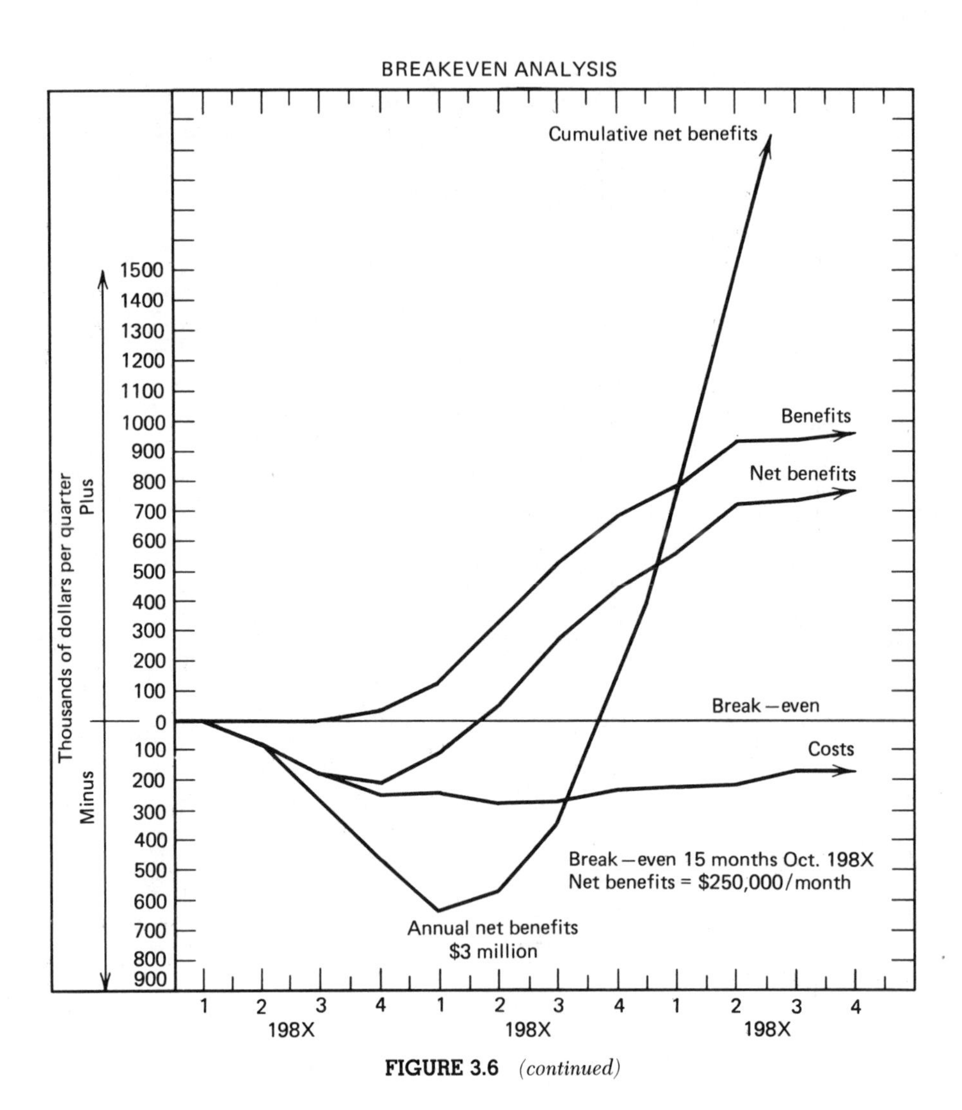

FIGURE 3.6 *(continued)*

Figure 3.7 is the study report implementation plan—the preface and a sample page from the plan. Every person assigned implementation responsibilities will have one or more pages similar to the one illustrated in Figure 3.7.

Implementation Plan
Preface

The following pages represent all tasks and goals that must be achieved.

The elements of work that occur within the first year of the project have been defined in greater detail (tasks) than those that occur later (goals). It is anticipated that in six to nine months, an internal planning session will be conducted by the Project Manager to regroup, redefine, and reschedule the remaining work based on the performance against this implementation plan.

The tasks on the following page are in sequence by:

- Task start date
- Individual
- Department
- Plant
- Division

FIGURE 3.7 Study report implementation plan section. *(Continued on next page)*

IMPLEMENTATION PLAN

Report A

Assigned Person: ______________________

Department: Inventory Control

Task Number	Planned Work	Planned Dates		Planned Time		
	Task Description	Start Date	Finish Date	Days Duration	Percentage of Time	Man days
99.93	Perform an initial audit of the item master (inventory accounting) data base.	12/23/81	1/5/82	5	100	5
99.99	Train appropriate user personnel in inventory polices and procedures.	12/22/81	1/5/82	5	100	5
99.100	Train appropriate user personnel in the use of the inventory accounting package.	1/5/82	1/7/82	3	300	9
99.83	Develop and document inventory accounting procedures not already developed for purchasing, receiving, dispatching, storing, and material planning.	1/5/82	2/2/82	20	50	10
99.49	Train dispatch planners on how to interface with the inventory accounting system.	1/7/82	1/11/82	3	100	3
99.103	Do a cost/benefit analysis for cycle counting.	1/11/82	1/25/82	10	N/A	2

FIGURE 3.7 *(continued)*

Figure 3.8 is the study report project management preface. It depicts the type of data relating to project management that will be included in the study report.

Project Management
Preface

The following pages on project management are provided as a recommendation for the project manager to track performance on the implementation. They include:

Plant project manager job description
Task tracking logic
Cost/benefit tracking logic
Sample tracking reports
Complete tracking reports
Blank tracking reports

FIGURE 3.8 Study report project management section.

Figure 3.9 is the study report appendix preface.

Appendix
Preface

This section includes all products that were produced at each step of the study methodology. It is included in the study report for two reasons:

1. To indicate to the reader the methodology employed to arrive at the tasks included in the implementation plan
2. To act as a guide for future replanning efforts

The contents portion of this section illustrates the sequence of the products as they were developed.

Physically large charts, such as the labor load-leveling chart, are not included in this text, but the data derived from these charts are included.

FIGURE 3.9 Study report appendix section.

This concludes your description of a final study report. Remember that you want to tailor the study report example (as well as all other examples) to your own specific needs.

At this point in actual presentation, you should have an audience that is interested in doing a study. You then cover the last presentation point—the study requirements—or "Here is what you have to do, Mr. Customer, to obtain all those benefits."

Figure 3.10 is a cover page for study requirements. It is a summary of all the things that we are going to talk about concerning requirements.

Study Requirements

A. Education
B. Available documentation
C. Facilities
D. Supplies
E. Study team members
F. A designated project manager
G. The executive commitment
H. An established work schedule
I. Agreeing to team rules

FIGURE 3.10 Study requirements—cover page.

Figure 3.11 is an example of a foil depicting available education relating to manufacturing. You should emphasize courses that your company offers and include schedule dates. The point to stress with this foil is that the study team should obtain as much education in the related study area as possible prior to the start of the study. Otherwise you will be conducting education with your team during the course of the study.

A. Education

Course Name

Manufacturing Executive Conference
Manufacturing Information Systems for Executives
Manufacturing Industry D.P. Directors Conference
Manufacturing Institute
Engineering Computing Conference
Aerospace Workshops
Manufacturing for Managers (VP and functional level)
Manufacturing Today
Customer Order Servicing—Implementation Planning
Engineering and Production Data Control—Implementation Planning
Inventory Planning and Forecasting—Implementation Planning
Material Planning and Reporting—Implementation Planning
Physical Distribution for Manufacturing Industry Executives
Shop Floor Concepts and Facilities
Purchasing—Implementation Planning
Receiving—Implementation Planning
MRP Video/Workbook Course
MRP Film Series

FIGURE 3.11 Study requirements—education.

Figure 3.12 is a foil concerning available documentation for study requirements. You should list the manuals and books that you think are applicable and, if they are available, list the sources and costs. Reference documentation should be available before the study starts. It actually is another source of education for your study team.

B. Available Documentation

Manual or Book

On-Line Customer Order Servicing—Executive Guide
Product Definition—Users Guide
Manufacturing Routings and Facilities—Users Guide
Inventory Management—Users Guide
Plant Monitoring and Control—Users Guide
Planning Systems Guide
Customer Order Servicing Application Description
On-Line Customer Order Servicing Application Guide
Designing A Net Change MRP System
MRP and Inventory Accuracy
Net Change MRP
Net Change MRP Implementation Considerations
Master Production Schedule Planning Guide
Plant Terminals—Production Summary
Data Base Guide

FIGURE 3.12 Study requirements—available documentation.

Figure 3.13 describes the physical facilities that are required for a study. This is provided by the company. It may be on-site or at an off-site location. Basically, you will want a room with no windows and space for twice the number of people that are on your study team.

C. Facilities

A Large Room

Space to stand and stretch
Maximum wall space with a minimum number of windows
Good ventilation

Tables

Comfortable seating facing one direction

A Source of Refreshment

Water, soft drinks, and coffee

Clean-up Service

Empty ash trays, trash cans, and so forth

FIGURE 3.13 Study requirements—facilities.

Figure 3.14 describes the supplies that will be required for the study. Although it is the responsibility of the company to provide these supplies, I used to bring my own tape and marking pens, because there were certain brands I preferred. Stay away from alcohol-based marking pens because they will soak through flip chart paper and leave marks on the wall.

D. Supplies

Two flip chart stands *with solid backs*
Ten plus pads of flip chart paper (lined with one-inch blue squares)
Water-based marking pens in various colors
Overhead foil projector
Projection screen (or a white wall)
One roll of masking tape
Ample pads and pencils for the group
One roll of transparent tape
Typing service (150 to 250 pages)
Access to a copy machine

FIGURE 3.14 Study requirements—supplies.

Figure 3.15 is a definition of who you want on your study team. The team should include from three to five people. Less than three gives you too narrow a perspective; more than five causes so much conversation that it extends the study duration. You obviously want key management personnel who understand the business on your team. Do not settle for the manager that the company can do without for six weeks. That manager does not have credibility, and, if you accept him, your final report will not have credibility. Part-time team members or late starters are also unacceptable. They were not there when you did the last step and, therefore, do not understand this step. As a result, you will spend a lot of time reexplaining what was done. The computer programmers and analysts are listed as unacceptable, because when you are concerned about reducing a million-dollar inventory by 10%, they will be concerned about the field size of the part number in the computer.

E. Study Team Members

1. *Size of Study Team*
   Three to five personnel

2. *Type of Personnel on Study Team*
   Material control manager
   Production control manager
   Master scheduler
   Inventory control manager
   Engineering manager
   Data processing manager

3. *Prerequisities for a Study Team Member*
   Understand manufacturing
   Understand the business
   Have top management credibility

4. *Unacceptable*
   Part-time team members
   Anyone joining the team after the study starts
   Programmers or analysts

FIGURE 3.15 Study requirements—study team members.

Figure 3.16 shows the criteria regarding a designated project manager. Sometimes when you use a foil like this, the company will respond with, "Well, we'll assign a team leader for the study now, and then we'll pick a project manager to control the implementation after the study is over." This is counterproductive. The two titles that are referred to should be the same person. If the company selects a project manager after the study is over, there is a good chance that they will select someone who was not on the study team. This person now has responsibility to install a plan that he or she did not assist in developing. It is not his. It is theirs and your chances of a successful implementation are decreased considerably.

F. A Designated Project Manager

1. *The Company should:*

   Select the project manager before the study starts and make him or her a member of the study team.
   Clearly define the project manager's responsibilities with a written job description.
   Adjust the Company organization so that the project manager reports as high as possible, within the organization.
   Communicate the above changes to the entire company staff.

2. *The project manager should:*

   Not be a new hire.
   Not be the DP manager.
   Be a person who understands the Company.
   Be a user such as the:
   - Material control manager
   - Master scheduler
   - Production control manager

FIGURE 3.16 Study requirements—a designated project manager.

Figure 3.17 is the executive commitment. Your audience (including your executive sponsor) should be in complete agreement with this foil. If you encounter any hesitation, regarding the expense, for example, be aware of a danger signal. Some companies will not spend money up front, regardless of the potential long-range benefits, and therefore the plan will never be installed.

G. The Executive Commitment

1. The people that I can least afford to do without will be on the study team.
2. I will stress applicable education and project management.
3. You will produce a realistic implementation plan, based on the defined key problem areas.
4. You will cost justify the system.
5. I will install the results of the study as scheduled by the implementation plan.
6. I will expend the monies for the schedule costs and also expect to receive the schedule benefits.

FIGURE 3.17 Study requirements—the executive commitment.

Figure 3.18 reflects the need to prepare an established work schedule. There are two key factors that will affect your work schedule.

Whether or not you will be traveling to the company and, if so, how far

The time that your study team will require off the study to continue running the business

I usually traveled to the study location and lived in a motel. As a result, I let the company select the daily start–stop times. It made no difference to me, since I was sitting in a motel room anyway. Before you take this foil off the projector, you should have some tentative agreement on the work schedule (point 4 on Figure 3.18) for this particular study.

H. An Established Work Schedule

1. *Hours per Day*
   6 to 9
   Start and stop times are by company choice
2. *Days per week*
   3½ to 5, depending on my travel constraints (all travel is done on Mondays and Fridays)
3. *Number of Weeks*
   Normally 6 to 8
   Possibly 5 to 9
4. *Our Daily Schedule*
   Monday: ______ to ______
   Mid-week: ______ to ______
   Friday: ______ to ______

FIGURE 3.18 Study requirements—an established work schedule.

Figure 3.19 is the final study requirement—agreeing to team rules. Although these rules appear firm (and you should be firm at this time), in actual practice you have to exercise a little tolerance. There are two rules, however, that should always remain firm! The first is that you are the session leader and you run the session. The second firm rule is that only one conversation is allowed at a time. If you back down on either of these, your study session will get out of control.

You should now have some ideas on how to prepare for your initial presentation. Remember that a little extra time spent tailoring your presentation to a particular company can make the difference between getting or not getting the study contract. These modifications can be simple things, like using the company's name instead of ABC Company and referring to specific problems or considerations of the particular company.

I. Agreeing to Team Rules

Each issue must have 100% team agreement.
Each team member has one vote (no pulling rank).
Only one conversation is permitted at a time.
Minimum interruptions, such as telephone calls (in or out), secretarial visits, fire fighting, and so forth.
If you leave the session and later return, *do not question any decisions made in your absence* as it is discourteous to make the group wait while you are briefed on what took place during your absence.
The session leader runs the session.

FIGURE 3.19 Study requirements—agreeing to team rules.

CHAPTER FOUR

HOW TO MAKE EFFECTIVE PRESENTATIONS

SOME DO'S AND DON'TS

Before we get into the do's and don'ts, let's review what you have done prior to your initial presentation.

You have received and reviewed a nomination form from the company.

You have received (and studied in depth) the company questionnaire from the company.

You have called the person who submitted the above two forms and set up an initial meeting.

You have met with a few of the company personnel and chatted about conducting a study—its benefits and its commitments.

You have gone on a plant tour and made notes about what you see being done right, as well as what is being done wrong.

You have set up a time and place for your initial presentation.

You have participated in the selection of the audience for your initial presentation.

You have tailored your presentation material to this particular company.

You have prepared handout material of your presentation for the intended audience, plus five extra copies for unexpected late comers.

You have made sure that items such as an overhead projector for foils (with an extra bulb), or flip chart stands, or whatever you are going to use will be available.

If you have just finished reading this list and concluded that you really do not have to do all of those things, then you have also concluded that you do not want to be a professional and are willing to accept significantly less income for a nonprofessional effort.

Now let's address some things that you should and should not do for the presentation. Some of these items may be obvious to you, but I have seen some presentations where these points were not observed.

1. If you are traveling to the company by plane, hand carry your presentation material. You can excuse a wrinkled suit due to lost baggage, but you cannot excuse deferring this key presentation because your materials were in the lost baggage.
2. Phone your executive sponsor the day before the presentation to ensure that everything is set up as you expected. Do this before you make the trip to the company site. At this point, you are operating on your money and the company, since it is not their money, may have forgotten to call and tell you that the meeting was deferred because a key executive was called out of town.
3. Get plenty of rest the night before the presentation. There is nothing worse than trying to give a pitch when you are tired or hung over. If company officials want to party with you, try to defer it until after the presentation.
4. Wear clean, freshly pressed, conservative clothes. This means a dark suit, a conservative tie, and polished shoes. It is immaterial that the company officials may dress in sport clothes. You are the outside professional consultant, so look like one. When the study is in progress, then it is the time for you to dress the way that your study team does.
5. Mixed opinions exist on wearing rings, tie tacks, and so on. One opinion is that if you wear diamond rings and tie tacks, then you must be a success, and everyone wants to associated with someone who is a success. Another opinion is that if you wear diamonds, then your rates are probably too high and maybe one of your competitors has lower fees. In general, your basic guideline is to continue to remain conservative in your dress.
6. Show up at least a half hour before the start of the meeting. This will give you time to check out the meeting room and chat with your executive sponsor. Ask him or her if there are any key issues that he or she would like to see covered (or avoided) during the meeting. Remember, the executive sponsor pays you, so play by his or her rules.
7. Try to meet everyone individually as they enter the meeting room. Ask what each person does and where their interests lie.
8. Start your presentation by giving some background on why you are here. The nomination was submitted by ______. The company questionnaire was prepared by ______, and ______. It looks like some of the key concerns

might be ______, ______, and ______. Your purpose here today is to describe a proven methodology, which can address those key concerns. And now go into your presentation.

9. While giving your presentation, be aware of these points. Do not put your hands in your pockets. Do not rattle your change. Do not smoke. Do not fold your arms across your chest. Do not put your hands behind your back. Have some water or coffee available for a dry throat. Use your hands for gestures to emphasize points that you are making in your pitch. Be enthusiastic. You must convey that this idea is the greatest thing ever; your potential customers will not get the idea all by themselves. And above all, do not do these things. Don't pick your nose! Don't scratch! Avoid using words like, "Ah," "Oh," and "Um."

If this is going to be your first presentation, you are going to get nervous. Some people will tell you that you'll get over that after you've done 10 or 20 pitches. Let me say that if you get over being nervous completely, you are on the downhill slide of your career. Sure, you should be confident. You do a job and do it well. But when you stop being nervous, you start being cocky. The impression that you convey is, "I'm so good at this that I really don't need you Mr. Customer." And guess what? With that attitude, you won't have "Mr. Customer."

AVOIDING PREJUDGED CONCLUSIONS

Initially, the customer told you that his key problem is a lack of master schedule planning. Knowing this, you ignore the company questionnaire. You merrily progress into your presentation, and maybe even make comments about what wonderful things that you have done on master schedule planning with other customers. You indicate that since this is a key area, it obviously is the first and major area that will be addressed by the study. Next come the questions from your audience:

> "I always thought that a production planning system should precede master schedule planning. Shouldn't we do that first?"
>
> "Since we have a poor order entry system, how can we do demand analysis which precedes even production planning?"
>
> "Heck, don't worry about order entry. We don't do forecasting either, which is the other function necessary for demand analysis. Does your study cover that?"
>
> "Ah, come on. You know that we ignore what Marketing wants and just build what we think that we can."
>
> "Sure, and that's why the warehouse is filled with finished goods that nobody wants. I suppose that a master schedule planning system is going to get rid of all that stock."

If your whole pitch has centered around master schedule planning and you got those questions at the conclusion of your pitch, you have lost. You didn't understand the problem. You assumed that your sponsor could correctly identify the problem.

What you should do is say that master schedule planning has been identified as a key concern. However, if it appears during the course of the study that there are some prerequisite systems, policies, or procedures that must be implemented to make master schedule planning effective, then obviously these will be addressed first in the implementation plan.

Be aware that usually a company will tell about the *key* thing that is wrong and should be corrected by the study. And usually they are wrong. Their "key thing" will be in your plan, but five years later, when all of the prerequisite problems are fixed.

This is an area where you have to be tactful and say, "I understand your key problem and of course it will be evaluated with all other problems that come to light during the problem definition phase. And of course, we will review the assigned problem priorities with you."

OBTAINING COMMITMENTS

By the time that you have finished your presentation, you should have "asked for the order."

"Mr. Customer, does it sound like this study methodology would be of benefit to you?"

If the answer is "No," then your response is, "Well, what are the points that concern you that I can change to make it acceptable for you?" And you negotiate.

If the answer is "Yes," then your next questions are:

1. Who are going to be the study team members?
2. What is the study start date?
3. Will the previously discussed working hours be acceptable to the team members?
4. Where (what physical location) will we be doing the study?
5. Can you give me the name of someone who will be responsible for setting up the study room with the supplies that I have indicated are needed?
6. Can you identify for me if there is a certain scope limitation to this study? For example, a specific plant or specific function within a plant?
7. Are there any goals that you have which should be achieved by the study team?
8. Would you consider distributing a study team memo (see Figure 4.1) to insure that all concerned personnel and their management are aware of what we are going to do and why?

You now have a study in an approved start-up mode. You are close to beginning your first study.

FIGURES

Memo to: (Each member of the planning session)
Subject: Study Schedule on How to Plan the Implementation of Manufacturing Information Systems.

We at (Name of company) will be conducting a planning session to define our direction in the (application) area. Those personnel addressed by this memo will be participants in the planning session.

The session will be conducted by (your name) of (name of company you represent). It will be conducted at (time) on (date) and will continue thereafter from (time) to (time), Monday through Friday, for approximately (number) weeks.

Please arrange your schedule accordingly.

Signed: (Top management executive sponsor)
Copies to: (The immediate manager of each of the participants to whom the memo was addressed)

FIGURE 4.1 Sample study team memo.

PART TWO

HOW TO LAUNCH THE STUDY

CHAPTER FIVE

GROUNDWORK: LAYING STRONG FOUNDATIONS

GUIDELINES FOR MEASURING PROGRESS: EVALUATING RESULTS OF A STUDY DAY

Everything that you have done so far and everything that you do in the course of a study for any specific customer should be written down as study notes or a study log. This means that if you make a phone contact to set up a meeting, you write down when you made the call, who you talked to, and the results of the call.

As you are doing the study, you should set aside some time each evening to update your log. Make your own personal evaluations regarding the following questions:

1. How did today go?
2. Was today productive?
3. Are we on the right track?
4. Are we on schedule?
5. Was there anything today that was outstanding?
6. Did anything happen today which might be the first sign of a potential problem?
7. What will we cover tomorrow?

Depending on the daily events, your notes for any one day may be a couple of lines or may cover a full page.

If you get into the habit of keeping a study log, you'll find it a very useful tool to:

Keep your manager aware of your progress (if you are not a private consultant)
Perform evaluations as to why Study A was better or worse than Study B
Use for follow-ups six months or a year after you have completed the study

It is not a good idea to reveal to the study team that you are maintaining a study log. If you tell them that it exists, then they will want to see it. If you know that it is going to be read by others (besides your own manager), then you are likely to be less candid than you normally would.

HOW AND WHY TO LOG TEAM ATTENDANCE

In addition to daily study notes, you should record daily team attendance in hours. Use a simple grid with team member names down the left side of the page and study days across the top. Then, as you progress through the study, enter for each day the number of hours that each team member (including yourself) was in attendance. Your attendance log could look something like the following:

Company: The ABC Company

	May						June	
	M	T	W	T	F	T	W	T
Team Member	23	24	25	26	27	31	1	2
You	4.0	8.0	8.0	8.0	4.0	4.0	8.0	
John Doe	4.0	7.0	6.5	8.0	4.0	8.0	8.0	
Bill Smith	4.0	8.0	8.0	7.0	0	4.0	8.0	
Jack Brown	2.0	8.0	7.0	8.0	4.0	4.0	8.0	
Phase	Problem Definition					General Design		
Customer total	10.0	23.0	21.5	23.0	8.0	12.0	24.0	

Your study team should be aware that you are keeping this attendance log. They should also know that it is available to the executive sponsor for his or her review upon demand. That little fact will tend to decrease a lack of attendance for miscellaneous or trivial reasons.

At the conclusion of your study, you will have a good analysis of:

1. What phase required most of your time?
2. Was any one person consistently absent?

This is a worthwhile log to maintain. You may need it to validate some specific point, perhaps only one study out of three, but it is a lot easier to build the log as you go along each day than try to reconstruct it five weeks into a study.

CHECKPOINTS—THE INTERIM REVIEWS

You may have noticed that in Figure 3.3, the sample study agenda for your initial presentation material, item 15 said "Conduct an Interim Review." The sample agenda only mentions an interim review once and you may well only do one. However, be prepared to do as many as necessary.

The purpose of interim reviews is to say, "Hey, Executive Sponsor, here's what we did, and here's what we found. Do you agree? Are we on the right track or did we sidestep since the last review? Do you approve of our proceeding on this path?"

The review is a checkpoint. It keeps your executive sponsor informed. I did one study that had no interim reviews, because the president came into the study room every day at 4:30, sat in the back of the room, listened, and read all the charts we had prepared that day. There was no point to schedule a review.

Conversely, sometimes you will be doing interim reviews to educate and sell your ideas to other company personnel. The executive sponsor who sets up an interim review with an audience of 30 managers is doing it to sell the study program to those 30 managers. He or she wants to build support by getting them involved while the study is still in progress. If you find yourself in this situation, recognize two things:

1. It will consume some extra time out of your study schedule to prepare the reviews and prepare for the questions that may be asked by 30 managers.
2. This type of an interim review is a 99% guarantee that the installation will be a success.

Let me point out that I personally never did an interim review. I know it would have been easier and quicker if I just did it, instead of:

Having my team decide if one or more of them should do the pitch and, if more than one team member was to do it, then who does what section.

Having the team frame the material that would be presented.

Having the team go through one to four practice runs to get a professional sound to the presentation.

My philosophy has been "If the company study team, which is building the implementation plan, is unwilling to stand up and talk to the company's own managers about what they have done, why should I, an outside consultant, do it."

Tell your team on day 1 that the interim reviews are their responsibility. Avoid any surprises.

Do reviews when you, the team, and your executive sponsor agree that they should be done. A study with a lot of "hot issues" should have more interim reviews than one with no "hot issues."

The content of an interim review should at least cover:

1. What has been done.
2. What key things have been found out.
3. What is the current overall plan.
4. What steps remain in the study.

One final reason for an interim review is to address the situation when things are going wrong or you are up against a blocking force that you cannot overcome.

For example, you are doing a study for a plant. They buy their data processing service from a central corporate data processing organization. You are doing the cost/benefit analysis in your study and you need to know how much DP is going to charge to run inventory accounting for the plant. DP tells you to figure out the number of terminal transactions and then multiply by 5 cents per transaction. You do that and proceed on your way through costing. DP calls and says that you should also estimate disk file space and add one cent per record. Okay, you do that, go back and rework all the cost extensions that you have done, and proceed on your way. DP calls and says that they just realized that you may want to do update transactions as well as inquiry transactions on the terminals. Inquiries are 5 cents, but updates are 15 cents. Okay, scrap everything that you have done and start again.

Obviously this scenario could go on forever. But you said you could do this study in five weeks. Four and a half weeks are gone. Do not tolerate this nonsense. Talk to your executive sponsor. Get an interim review scheduled. Discuss the issue during the review. Get resolutions to problems like these quickly. Let your executive sponsor handle the politics.

In short, interim reviews are of benefit to you, although they cost time, so use them when you need them.

PREPLAN THE FINAL REPORT—AND SAVE TIME LATER

You can build your final study report in one of two ways.

1. Wait until the end of the study and spend an extra week just working on the final report.
2. Build the report as you progress through the study so that at the end you only have some final editing and copying left to do.

Obviously, I select the second choice above. I will not expend an extra week on a routine, mechanical task like a study report.

You have seen the sample final study report in Chapter 3. It consisted of these major sections.

Cover page
Preface
Management summary (with key summary charts)
Implementation plan
Project management
Appendix

The cover page is easy. Do it last. The preface can almost be used as is. The management summary should be done next to last, but you can frame it as you progress through the study. Project management material is covered in Chapter 11. The appendix includes everything else that was written on a flip chart that does not appear elsewhere in the report. The data in the appendix should appear in the sequence in which it was generated.

Look at Figure 5.1. The agenda is the first item in the appendix, because one of the things that I do when the study team first assembles is to develop a tentative agenda (it normally changes, so I also tell the team it probably will change). The material content then progresses just as the flip chart pages were developed during the study.

As I'm doing a study, I know that I want the agenda in the appendix. Therefore when I make the agenda flip chart page, I label it "A. AGENDA." The next item is obviously labeled "B." Then about once a week, I'll take down all of the flip charts that I think can be spared and send them in for typing. I then edit and assemble the typed material for the final report.

Prior to a final edit of the study report by the study team, try to get a complete typing pass and eliminate as many spelling errors and typos as possible.

The average study report takes three typing passes. And it's a chore if it is done on a standard typewriter. A better choice, if available, is a word processor or a magnetic card typewriter.

With a little bit of practice, the whole process of preplanning and framing your final report will become almost automatic.

Appendix Contents

Problem Definition Phase

A. Agenda
B. Interview schedule
C. Defined detail problems
D. Sorted detail problems by functional area
E. Major problems
F. Prioritized major problems
G. Checkpoint review #1
H. The study team objective

General Application Design Phase

I. Current operational logic
   1. Overview flow
   2. Estimating process
   3. Set up the master schedule
   4. Set up the drawing and detail schedules
   5. Plan long lead time items
   6. Review specifications
   7. Establish engineering files
   8. Obtain approvals
   9. Develop historical data
   10. Miscellaneous engineer activity
   11. Identify owner furnished materials
   12. Shop bill generation
   13. Shop bill and drawing revision
   14. Design change notices
   15. Engineering change requests
   16. Work order issue
   17. Work order changes
   18. Inventory control/warehouse
   19. Hand requisition types
   20. Credit back to the warehouse
   21. Warehouse receiving
   22. Expediting
   23. Physical inventory accounting
   24. Production planning

FIGURE 5.1 Partial appendix contents.

J. Selection of systems to be planned
K. Work order relationships
L. Product structure relationships
M. Design operational logic
    1. Master network relationships
    2. Sample network of side sections
    3. Single level work order bill of material
    4. Sample drawing
    5. Shop bill
    6. Sample work order material requirements
    7. Sample screen of purchase requirements
O. Goals necessary to overcome major problems
P. Goal dependency relationships
Q. Rough cut goal duration and date estimates
R. Checkpoint review #2

(The remaining two phases have not been included in the example)

FIGURE 5.1 *(continued)*

CHAPTER SIX

START-UP ACTIVITIES

REVIEW OF THE EXECUTIVE SPONSOR'S PRESENTATION—WHY THE REVIEW IS NEEDED

It is very likely that, although you now have your study team assembled, you may never have previously met some of the team members. Not only do you not know who they are, but neither do they know who you are. You need a little get-acquainted time.

To start with, pass a sheet of paper around the room that has these headings on it:

	Your Name	Your Title	Major Responsibilities
1.			
2.			
3.			

This simple information sheet serves several purposes. You now have the correct spelling of every team member's name. If you keep track of the sequence that your piece of paper was passed throughout the room, you know that the guy in the third seat back on the right is Bill. It is more professional than saying, "Hey, what's your name?" The Major Responsibilities section provides you with some insight as to where your support is going to come from as you progress through each of the study phases.

Next, spend a few minutes describing yourself. What is your background? Why are you qualified to do this study? How did the study get started?

Now ask each of the team members to do what you just did. What is their background? Why do they think they were picked for the team? What do they hope to get out of the study?

In order to obtain this study contract, you did an initial presentation for the executive sponsor. It is worth your time to repeat the entire presentation at this time, even if the entire study team was present at the initial presentation.

On the first pass, the material in your presentation was new. The executive sponsor was present, and your team may have been reluctant to ask questions. Now you can take twice as long for the presentation. Answer all the questions as you do the presentation. Make sure that the team understands the type of Final Report that will be produced. Make sure they understand and agree to the team rules. Reemphasize the key rules:

1. You are the session leader and you run the session.
2. Only one conversation is allowed at a time.

Now go back to the Sample Agenda in your presentation foils. You now know something about this company and its problems. The study team now knows something about the methodology. On a flip chart, build your first cut at a tentative agenda. Move step by step through your foil example and decide if more or less needs to be done on each step. This is a team effort and requires everyone's participation. Hang the flip chart (agenda) on the wall where everyone can refer to it and check their progress. By the way, ordinary masking tape works very well for hanging flip charts on a wall.

When you have completed your agenda, ask your team to determine when interim reviews should take place. Add those to your agenda and assign your project manager the responsibility of setting those reviews up with the executive sponsor.

DOING A COMPANY ANALYSIS QUESTIONNAIRE

A sample Company Questionnaire is included in the appendix for your use. You should either modify it or construct your own for your specific needs. If you create your own questionnaire, remember to keep it simple. Try to use only multiple-choice and true–false questions.

You should use a company questionnaire when you set up this study. If you do, the questionnaire is either completed by one person, or several people each complete sections of it. Now you should use it as a start-up activity.

Put aside your completed questionnaire. Make foils of a blank question questionnaire and put them on the overhead projector. On each question, ask for a vote. With a transparency marking pen, write the vote on the foil. For example, for question 1: 2 votes true, 3 votes false. This exercise will do several things for you.

1. A lot of discussion will be generated.
2. The team will start to realize that no one person knows all of the answers.
3. You will be able to add questions of your own, expressing interest in the company and gaining knowledge about company operations.
4. You will start building a strong rapport with your team.

After your team has gone through all of the questions on the questionnaire, take a foil pen of a different color and mark in the answers that you received when you initially submitted the questionnaire. You will rapidly realize that those areas with the most indecision as to an accurate answer are areas that should have further investigation during the Problem Definition phase. You might want to interview additional people who are involved in those areas of indecision.

Keep the company questionnaire foils that you and the team have developed. At the end of the study, you may have some slack periods, like when you are waiting for typing or copying. During one of these periods, review your questionnaire foils and see if the answers change now that your team has gone through an in-depth analysis of the company. The results are often both interesting and amusing.

DEFINING FUNCTIONAL RELATIONSHIPS

Everyone has the normal feeling of being different from the next person. The purpose of this step is to display to all members of your team that they really have the same functional interrelationships as other, similar companies, and, therefore, some similar solutions might be applicable.

You could start with a blank flip chart and build a function (not department) relationship chart. I start with a basic chart (Figure 6.1), which I display as a foil, and build from there.

For example, maybe your team wants to break the master production schedule planning block into demand analysis, production planning, and master schedule planning. That's okay—do it. Figure 6.1 is just a guide. What you want to determine is how your team views the functional relationships in this particular company.

At this point, some problem areas may come to light. For example, your team says that purchasing is driven from master schedule planning. You say, "That's not too logical. Why isn't purchasing driven from material requirements planning?" Their answers are:

Well, purchase items are expensive.

We have long lead times on purchase items.

We do it on a general level, because we only have a few buyers.

We don't want to confuse shop orders with purchase orders.

All of those answers are less than the whole truth, but eventually you get the real answer.

Our material requirements planning system is so bad that we can't trust it.

You have identified a key functional problem.

Keep this flip chart. In fact, redraw it so that it is nice and neat. You'll probably use it in your management reviews and in your final report to illustrate the interrelationships "before" and "after."

FIGURES

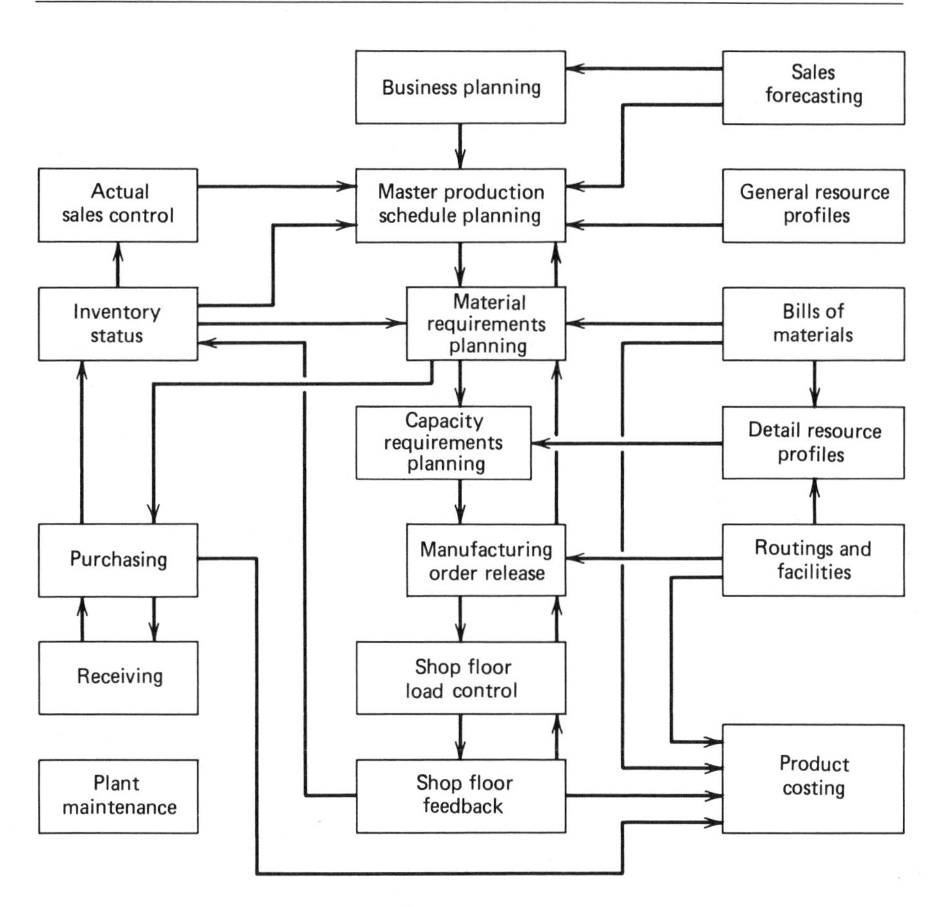

FIGURE 6.1 Example of manufacturing functions and their relationship.

PART THREE

CONDUCTING THE STUDY

The Four Phases

CHAPTER SEVEN

PHASE I: PROBLEM DEFINITION

PURPOSE OF PROBLEM DEFINITION

The reason that you are doing the problem definition phase is to find out the real problems in the company. You will be interviewing people ranging in function from a machine operator to the president. You will uncover a lot of detail problems during your interviews, and many of those problems will be mentioned recurrently.

You'll sift those detail problems down into the functional areas that you defined in Chapter 6. Then you'll analyze the set of detail problems in each functional area and decide if the set represents one or more major problems. You will define those major problems.

For example, you may interview 15 people and document 295 detail problems. You sift those down into 14 functional areas and define 27 major problems.

You now prioritize the 27 major problems and review the results with your executive sponsor.

Do you agree that these are the major problems?

Do you agree with the priorities that were assigned by the team?

Which problems do you think that we should address; the top priority problem only or maybe the top three priority problems?

This process sets your direction for the balance of the study, so now you can establish the objectives of your study team. That is the purpose of problem definition—to establish the objectives of the study team.

BRAINSTORMING FOR INTERVIEW ISSUES AND PERSONNEL

The issues that you want to explore during the interviews and the people that you want to interview obviously go hand in hand.

When you constructed the Functional Relationship chart in Chapter 6, you may have gained some insight as to areas of concern. Also, when you reviewed the company questionnaire with your team, some potential problem areas may have appeared.

At this time in the study, refer to the above two documents and have your team identify those issues that they feel are the most important. List them on a flip chart. Then brainstorm and list any other potential issues that may come to mind. By now you probably will have identified the key issues.

Have your team prioritize the issues by vote (this does not require 100% agreement—majority wins). Then identify those people to be interviewed that are the most capable of addressing the key issue. Repeat the process for all remaining issues.

Now you should have a list of names that will cover the probable key issues of the company. Look for gaps. For example, you may have top management and direct labor, but no middle management. Fill those gaps. Have representation from every level in the company.

HOW TO SET UP AND CONDUCT INTERVIEWS

You now have a list of people that you want to interview, and you probably have too many. A practical maximum is 15, but it is normally difficult to hold it to that for many reasons.

Let us assume that your company has three expediters. You have identified one that you want to interview. This is a poor idea. You had better interview all three or none, or you will have initiated hostility against your plan before you even have it defined.

Recognize that when you interview over 15 people, two things happen:

1. The amount of redundancy on detail problems increases significantly.
2. It just takes a lot more time for very little benefit.

You may set up a tentative interview schedule with one or two people per interview. Two is okay only if they both do the same job. Each interview should be scheduled to last 20 minutes. The interviews should be scheduled 30 minutes apart.

It is the responsibility of your project manager to contact each of the personnel to be interviewed, establish the interview time, and make sure that they understand that they are to come to the planning study room. Your interview schedule might look like the example in Figure 7.1

As each person comes into the room, every effort should be made to display that the team is seriously interested in the opinions of the person being interviewed. This is not an inquisition. It is a search for information. Except for questions at the closing of the interview, all conversation should be between the interviewee and the project manager.

To start off each interview, the project manager should spend two minutes explaining why the planning team was assembled and what they hope to accomplish. The project manager may have already conveyed that information via memo (a good idea if time permits) when he or she was setting up the interviews, but it is a good idea to repeat it at this time.

The project manager then explains that this planning group has only two basic questions.

1. What problems do you have in doing your job?
2. What do you feel you need in order to do your job better?

As the person being interviewed starts defining problems, list them on a flip chart. Do not put the individual's name on the problems, or he or she may be reluctant to tell you about key issues. List all problems as complete sentences. Start with Problem Number 1 and progress sequentially until the interviews are over. List *all* problems that each person tells you, no matter how redundant the problem is. If you start saying "Yeah, we heard that before," and don't write it down, then the person thinks that you are not interested in his or her input.

Figure 7.2 is an example of a partial problem list that you might develop. Note that every problem statement is a complete sentence and can be understood when read outside of the total list.

At the conclusion of this step, I have seen the number of defined problems range from as low as 60 to over 500. On the average, you should have 200 to 300 problems.

HOW TO GROUP DETAIL PROBLEMS BY COMPANY FUNCTIONAL AREAS

Let us assume that during the interview cycle you identify 325 problems. It is difficult to identify any major issues when you are looking at a list of 325 problems. You have to sort those detail problems down to some major grouping that is manageable.

Earlier you constructed a chart of functional relationships for this company. You may have identified 15 to 20 different functions. Now list those same functions on a flip chart. Leave a few inches of space between each function.

Now look at your list of detail problems and ask your team to tell you the *single* functional area under which Problem 1 best fits. The problem may relate to several functions but select the single one that is most applicable. Write the problem number under the function and repeat the process until you have grouped all of the detail problems by functions.

As you are going through this process, you are very likely to create some new functional categories that are not really functions, such as:

Policies and procedures
Education
Communications
Management concerns

When you have completed your grouping, you will have a flip chart that looks something like Figure 7.3.

HOW TO RESTATE GROUPS OF DETAIL PROBLEMS AS MAJOR PROBLEMS

This is best performed by team individuals and then refined by total team review.

Explain to your team that what you want is one or more statements that summarize the meaning of the detail problems within a functional area. These summarizations are called major problems. Each team member selects one or more functional areas. The team members work on the major problems individually. They may have one or more major problems defined for each functional area. They should not, however, just list the detail problems.

As team members complete major problem statements, have them transcribe the statements onto flip charts.

When the entire team has completed their tasks, review all of the major problem statements with the team. Here are some checkpoint questions.

1. Were all of the detail problems covered in the functional area?
2. Were complete sentences used?
3. Could a non study team member read the major problem and clearly understand its meaning?

Figure 7.4 is an example of some major problem statements.

PRIORITIZING MAJOR PROBLEMS

You now have a list of major problems, which in quantity should be about 10% of the number of detail problems that you developed during the interviews.

Let us assume that you finished with 32 major problems. What you would like to do now is to prioritize those major problems. With a quantity like 32, it is still difficult to look at any one of the 32 problems and judge its priority in relation to the other 31. To simplify the prioritization process, we use a two-step technique.

Set up a flip chart as in Figure 7.5. The Probable Resolution Value is the value to the company (high, medium, or low) of resolving a particular major problem. Have your team tell you, for each major problem, what the resolution value would probably be. Your flip chart now looks like Figure 7.6.

Now, instead of 32 major problems to prioritize, you have three categories that can be prioritized individually. Look at the 14 problems with the highest Probable Resolution Values. Of the 14, ask your team to tell you which one is the highest. Then the next highest, and so on. You chart now looks similar to Figure 7.7.

Repeat the process for the medium and low resolution value problems. Your completed prioritization chart should be similar to Figure 7.8.

DEFINING THE STUDY TEAM'S OBJECTIVES IN RESPONSE TO MAJOR PROBLEMS

You could have defined the objectives of the study team when you first started the study. If you did, you'd probably be changing those objectives right now, based on what you have learned through Problem Definition. If you define your team objectives at this point in the study, you only have to do it once.

The team objective(s) should be framed as one or two paragraphs. Consideration should be given to these questions.

1. Why are we (the study team) here?
2. What guidance (direction) has been provided by our executive sponsor?
3. What major areas of the business should we address?
4. Which manufacturing functions should be specifically addressed because of their importance?
5. What is it that we, the team, intend to produce as a result of this effort?

Each team member should make an attempt at stating the team objective. Have all team members read their statements to the rest of the team. Let the team select the best objective(s). Write the study team objective(s) on a flip chart and then fine-tune it.

This is an important step. It unifies your study team into a single company direction. It also is a statement to the rest of the company defining your direction.

You may want to do an interim review at this point. If not, you should at least do an informal review with the executive sponsor to insure that your defined direction concurs with the sponsor's wishes.

FIGURES

Date: 2/1/XX			
Time	No.	Name	Position
0900	1.	S. Hinkleman	President
0930	2.	B. Jones	VP Manufacturing
1000	3.	C. Doe	VP Finance
1030	4.	A. Kaminsky	VP Marketing
1100	5.	B. Johnson	Plant Manager
1130	6.	J. Brown	Engineering Manager
1300	7.	W. Ray	Purchasing Manager
1330	8.	J. Stewart	Shop Foreman
1400	9.	B. Black	Project Engineer
1430	10.	T. Burns	Personnel Manager
1500	11.	L. Baker	Farm Out Coordinator
1530	12.	C. Carter	Expediter
1600	13.	E. Charles	Master Scheduler
1630	14.	D. Taylor	Milling Machine Operator
1700	15.	R. Kay	Department Supervisor

FIGURE 7.1 Sample interview schedule.

1. Our inventory transactions are not well controlled due to timing.
2. Parts counting is not done accurately, since we do not use load cells for small parts.
3. We cannot get the right parts to the right place at the right time.
4. Our department responsibilities are not clearly defined.
5. We do not ship on the delivery dates that we quote.
6. We do not have effective security.
7. We have a significant part numbering system which is no longer effective.
8. Inventory status is 24 to 48 hours old.
9. Our labor reporting is inaccurate.
10. We need standardized bills of material.
11. We have too much paperwork.
12. We have trouble in defining and discontinuing obsolete parts.
13. It takes too much time to process a customer order.

FIGURE 7.2 Sample problem list.

A. Business Planning
   9, 18, 19, 20, 36, 37, 41, 44, 45, 58, 72, 75, 76
B. Forecasting and Demand Analysis
   24, 28, 29, 95, 103, 104, 150, 166, 175, 176, 192
C. Production Planning
   3, 21, 23, 26, 30, 31, 39, 61, 64, 71, 80, 85, 90
D. Master Schedule Planning
   4, 46, 53, 57, 73, 74, 77, 107, 110, 120, 121, 127
E. Bills of Material
   40, 105, 118, 119, 121, 142, 156, 327, 240, 260
F. Order Entry and Inquiry
   7, 15, 17, 33, 51, 55, 86, 87, 89, 111, 129, 131
G. Inventory Accounting
   14, 16, 47, 56, 135, 138, 151, 161, 163, 189
H. Material Requirements Planning
   1, 2, 6, 22, 27, 63, 82, 98, 145, 157, 158
I. Purchasing
   34, 42, 59, 91, 92, 93, 97, 101, 106, 109, 112

FIGURE 7.3 Grouping detail problems by functional areas.

A. Shop Floor Control and Feedback
   1. Our present capability to monitor and control shop flow is severely hampered by our lack of documented moves. This results in the loss of shop order identity, order location, and the ability to purge old orders or perform job assignment.
   2. Quality Control personnel are unable to perform an effective job due to a lack of equipment, manpower, procedures, and training. This causes high rework costs, scrap, and poor quality on shipped parts and equipment.
B. Capacity Requirements Planning
   3. The scheduling of orders on the floor is being done by expediters and foremen. Consideration of the underloading and overloading of other departments is ignored. Requirements of tooling, tooling maintenance, foundry delivery, manpower, and machine capacities are not considered. Orders are released too far ahead, without knowing priorities.

FIGURE 7.4 Sample major problem statements.

Major Problem	Probable Resolution Value			Priority	Major Problem	Probable Resolution Value			Priority
	High	Medium	Low			High	Medium	Low	
1					18				
2					19				
3					20				
4					21				
5					22				
6					23				
7					24				
8					25				
9					26				
10					27				
11					28				
12					29				
13					30				
14					31				
15					32				
16									
17									

FIGURE 7.5 Basic chart design for prioritizing major problems.

Major Problem	Probable Resolution Value			Priority	Major Problem	Probable Resolution Value			Priority
	High	Medium	Low			High	Medium	Low	
1	X				18	X			
2	X				19		X		
3			X		20	X			
4		X			21	X			
5			X		22	X			
6			X		23	X			
7		X			24		X		
8		X			25	X			
9	X				26	X			
10		X			27	X			
11		X			28			X	
12		X			29			X	
13			X		30			X	
14	X				31			X	
15			X		32			X	
16	X				Totals	14	8	10	
17	X								

FIGURE 7.6 Resolution values defined for major problems.

Major Problem	Probable Resolution Value			Priority	Major Problem	Probable Resolution Value			Priority
	High	Medium	Low			High	Medium	Low	
1	X			6	18	X			5
2	X			7	19		X		
3			X		20	X			8
4		X			21	X			9
5			X		22	X			14
6			X		23	X			12
7		X			24		X		
8		X			25	X			3
9	X			10	26	X			1
10		X			27	X			2
11		X			28			X	
12		X			29			X	
13			X		30			X	
14	X			13	31			X	
15			X		32			X	
16	X			11	Totals	14	8	10	
17	X			4					

FIGURE 7.7 Priorities assigned for high resolution value major problems.

Major Problem	Probable Resolution Value			Priority	Major Problem	Probable Resolution Value			Priority
	High	Medium	Low			High	Medium	Low	
1	X			6	18	X			5
2	X			7	19		X		21
3			X	28	20	X			8
4		X		19	21	X			9
5			X	25	22	X			14
6			X	26	23	X			12
7		X		20	24		X		17
8		X		15	25	X			3
9	X			10	26	X			1
10		X		18	27	X			2
11		X		16	28			X	24
12		X		22	29			X	30
13			X	29	30			X	23
14	X			13	31			X	27
15			X	32	32			X	31
16	X			11	Totals	14	8	10	
17	X			4					

FIGURE 7.8 Major problems completely prioritized.

CHAPTER EIGHT

PHASE II: GENERAL APPLICATION DESIGN

PURPOSE/OBJECTIVE OF THIS PHASE

Your purpose during this phase is to produce a General Design that will resolve the key major problem(s).

You have defined the relationship of the business functions for the company. You have also gone through a problem definition phase and identified the key major problems. You are not going to solve all of the major problems with a single cut of an application design. The most that you can hope for is to pick the one, two, or three key problem areas and address those.

Suppose that during the problem definition phase, you determined that the top three (highest priority) major problems were as follows:

Priority	Major Problem
1.	It is difficult to do a reasonable job of production planning, because our forecast data are unrealistic.
2.	Because we have no provisions for scheduling plant maintenance, we incur a lot of remedial maintenance (resulting in unscheduled machine downtime).
3.	Move times of shop orders between departments are vague, resulting in inaccurate lead times and poor material planning.

These three key problems are unrelated. They are obviously crucial to the business, and all three need to be solved, but they are not related to a single application design. Bite off one piece at a time. If your study team designated the unrealistic forecast data as the number one priority, then that is the one that you should attack.

Now remember that this is called a "general" application design. Keep it general. Stay away from the data processing bit and byte level. This is not always so easy to do because there frequently are individuals associated with your study who try to drive you down to the most minute aspect of your design. They do this for one of two reasons:

1. They cannot see (or care to see) the overall approach. They live in a world of detail. Their typical comment is, "I don't care about the overall forecasting function. Tell me how I can take the actual sales data for Item 123 and blend them to the forecast data in Period 3."
2. They may be ultimately responsible for the detail design. If they can get you to do more now, then they do less later. Their comment is, "That's too general. Nobody can write a program from that." They're right, and that's why it's called a general design. The implementation plan will define the tasks for the complete detail design.

Do not, however, get carried away with keeping it general. If your design for forecasting consists of a single statement—"We will build a forecasting system"—then you are much too general. You need enough information in your design so that the major steps can be understood.

There is one more point to be aware of. Suppose that you want to do a General Design of the key major problem, but you can't. Too many other things seem to get in the way. Frequently the key problem is really crucial but it's not logically the first one to address.

Let's consider an example of this situation. You previously constructed a function relationship chart, and it may have looked like Figure 8.1.

Your key major problem may have been "The current material requirements planning system is not effective." However, during the problem definition phase, you may have also learned the following:

1. We do a poor job of educating our people on how the Company runs.
2. We suspect that many of our purchasing and manufacturing lead times are wrong.
3. We question the planned order dates and quantities from master schedule planning.
4. We run MRP monthly.
5. We think our inventory accuracy is about 56%.

If your problem was MRP and you know that the inventory accuracy (for example) was 56%, then what is the problem for which you must design a solution? It's not MRP. Even if you put in the most sophisticated MRP system possible, you will not fix the problem. No MRP system will work if the inventory accuracy problem is not solved first. Therefore, you need to know not only what your group thinks the priority problem is but also how it relates to other functional areas.

Your intent is to produce a general application design—but not too general. You are going to concentrate on the key major problem, or problems, if the top priority two or three are related. And, finally, you are going to be sure that there are no prerequisites to your key problem(s) which must be resolved first.

DEFINING THE CURRENT OPERATIONAL TECHNIQUES

Operational logic is a description of how the various aspects of the business interrelate. There are two kinds of operational logic—current and desired. Current describes how the business runs today. Desired describes how you would like the business to run in order to overcome the key major problem(s).

Some managers feel that defining the current logic is a waste of time. They say, "Sure we've got problems today. We know that. Let's get on with how we want it (the system) to work and quit wasting time on further definition of how we do things today." Okay, let's say you get conned into agreeing to that. The next phase is implementation planning. You have defined where you want to go. You have not defined where you are at. How will you define the first task necessary to get your from "here" to "there" if you don't know where "here" is?

For example, you might want to design an inventory accounting system. However, your company already has an inventory accounting system. You obviously have to do two things:

1. Identify how the function is performed today.
2. Identify how the function would be performed with a new system.

This will allow you to evaluate the "gap" between today's system and the new system and therefore figure out what steps have to be performed to get from here to there.

There are several ways that you can document the current operational logic as well as the desired operational logic. A simple way is to use straight text, as opposed to diagrams.

Suppose that you have an estimating process that is required for the generation of a contract that you wanted to document. If you used text, your steps might look like this:

1. The customer has an idea and submits a request for a price quote.
2. The company does an estimate and returns it to the customer.
3. The customer signs a contract.

Or does the customer sign a contract? Maybe he or she makes changes and the Company has to reestimate. Maybe the reestimate cycle is zero times or 10 times, depending on the customer. This is where straight text falls apart as a documentation procedure except for very simple cases. Using text, it is difficult to illustrate (describe) alternate paths.

A more successful approach to the documentation of operational logic is with a flow chart. It does not have to be pretty, artistic, or done by a professional graphic artist. It just has to be accurate. The estimating example is illustrated in the top portion of Figure 8.2.

As you look through Figure 8.2, think how—if you were using text—you would treat the following cases:

1. There may be desired options.
2. There may be long lead items (items with a long procurement cycle).

Figure 8.2 was a general overview. You should select key areas from your general overviews and make more detailed flow charts. For example, in your general overview you may have had a logic block that said "Obtain material from warehouse." If it had looked like (in the problem definition phase) inventory accuracy was a problem, you may want to expand this logic block, the flow of which might look like Figure 8.3.

As you review Figure 8.3, you become aware of where some of the inventory control/accuracy problems might exist.

There are times, when you want to depict a concern, that you will want to add lists of data to your charts. For example, in your flow charts, it seems like a lot of things are routed to and originated by a line foreman. You suspect that he is floundering and out of control. You make a list like that in Figure 8.4 to identify just what documentation he manages.

When you complete this segment of the study, you should have three to ten general overview flip charts plus another 12 charts which provide more detail on specific areas of concern. Leave these charts taped to the wall of your study room because you'll need them for reference.

DEFINING THE DESIRED OPERATIONAL TECHNIQUES TO SOLVE KEY PROBLEMS

In this section you will define how you would like the business to run. For documentation you should use the same techniques that you applied in the current operational logic, because you will be comparing the two sets of logic in the next section also.

So far you have:

Identified the key major problem(s)

Built a chart showing the interrelationships of business functions (Figure 8.1)

Defined the current operational logic

What you have to do at this point is to get your group to agree on how much of what you have uncovered (documented) will be addressed in your design logic. There is no nice, clean, step-by-step procedure to do this. You have to discuss your way through this decision process. Start with your functions chart (Figure 8.1). The key problem was material requirements planning. This function is fed data by Inventory Status, Master Production Schedule Planning, and Bills of Material. When you determined the current logic, did any of these three functions have severe problems? Your group says that both inventory status and bills of material have problems that need correcting before MRP will work well. Okay, what else? Review the other functions. If you find some other functions with problems, ask your group about the severity of the problems in those other functions, as related to your key problems. If they aren't crucial, keep them on the back burner. If you have time, you may want to cover them after your key problems of inventory status, MRP, and bills of material.

At this point, you may want to do an informal review with your executive sponsor. Show him or her what you've found and what you intend to address with a solution. Summarize those other "not so important" areas. Find out what he or she thinks is important. This is a good checkpoint, but beware. Be prepared to defend your conclusions to date. If you just casually explain that inventory status has to be fixed before MRP, the sponsor may say that he or she thinks some other area, such as purchasing, is really key. Well, with the steps that you've gone through, you know purchasing looks fine for now. Why did your sponsor narrow in on purchasing? Who knows? Maybe he or she doesn't like the purchasing manager. It doesn't matter, because once you've let the sponsor state his or her preference, he or she has to stick with it or lose face. Do not get yourself into this box. While you have the floor during your informal review, explain the steps (and logic) that took you to your conclusions. If he or she understands that logic, then the only possible responses are:

1. Do inventory status first.
2. Do bills of material first.
3. Do both inventory status and bills of material.
4. Do inventory status, bills of material, and those other areas that you put on the back burner.

No one could logically say "Do Purchasing." If that is the executive sponsor's response, you should terminate the study. If you continue (and do address Purchasing), you will be designing a solution to a nonexistent problem. You will be doomed to failure, so you are better off to get out now and not waste your time.

So far you have identified the key areas that you will address: inventory status, MRP, and bills of material. You've shown your findings to your executive sponsor. He or she understood and approved. You're going to use the same documentation techniques that you used with Current Operational Logic.

When you start to build flowcharts on desired logic, you will be faced with comments like these by your group:

1. We can't do much, because we don't have the budget.
2. Let's really blue sky it, because all things are possible.
3. Why don't we just identify some quick fixes to get us out of trouble?
4. The whole thing is hopeless.

Your comments should be something like:

1. At this point, we should not worry about what the solution will cost. Remember that we are addressing the key areas (to which our Executive Sponsor agrees) and we will find cost-effective solutions.
2. Sure, all things are possible, but not always practical. We could replace everyone in the Company with robots, but the executives (and employees) may not jump at the idea. We have to be realistic. What we plan for has to be based on technology that exists today. No betting on the future.
3. We do however, have to look at a total system approach. We should not build an Inventory Requisition Handling System with total disregard to the other aspects of Inventory Accounting and the interfaces with other functional areas. As such, our implementation plan may span three years, five years, or even ten years. That's okay. Our goal should be to plan for the total picture for our key areas as we see them today.
4. If any of you truly feel that the effort is hopeless, then you do not understand the process that we are going through—which works—and you do not have to continue as a member of this group.

As in current operational logic, you will have general design charts and then detail charts for specific areas of concern. Figure 8.2 depicted the current logic for estimating a contract and getting the order into the system. Figure 8.5 might be one chart (of several) that reflects the general desired logic to apply to these functions.

Support your general level charts with detail charts where additional definition is required. Note that when you finish these charts, you are not necessarily done with this step. Throughout the remainder of the study, you may want to come back and develop more detail.

When you complete this segment of the study, you should have five to 15 general overview flip charts, plus another 10 to 20 detail charts. Leave these charts on the wall with the charts that you prepared for the current operational logic.

THE DEFINITION OF GOALS TO ACHIEVE THE DESIRED OPERATIONAL TECHNIQUES

The purpose of this section is to review the current operational logic and identify the major elements of work necessary to get from "current" to "desired." These major elements of work are called goals. Later on, we use an element of work called a task. The distinction between these two work elements is as follows:

Goal. An element of work that is achievable and measurable. (Duration is one month to one year.)

Task. One of the steps required to accomplish a goal to which a realistic duration can be applied and a responsible person can be assigned. (Duration is one to ten days.)

If you explain these two terms to your group, you are likely to get this reaction: "Why do we have to bother with the goal step? Why don't we just define tasks?"

First off, remember that you are building a plan that could take three to ten years to implement. If you only defined tasks (and skipped goals), how realistic do you think your tasks would be in five years? Also, a task gets assigned to a specific person. Do you think that person will be in his current position in five years? The answer to these questions is obvious.

A more practical approach is to define goals for the entire plan. Then establish a time fence, say of about six to nine months. All goals that fall between now and the end of the time fence should be subdivided into tasks. This gives you a plan that is very detailed at the front end and more general the further out that it goes. Recognize that you are going to have to revise your plan periodically. You'll want to, because things change. Your first revision should take place at the end of your time fence. After that, it should be done once a year.

The result of this process is that you will have a "Rolling Plan." It has front-end detail with individual responsibilities, plus the long-range goals, and it is updated every year.

A goal is a major element of work. You will need individual thinking as to what goals are required. Your desired and current operational logic charts are on the wall. Have each member of your group take out pencil and paper and individually write down the goals that they think are necessary to get from the Current Logic to the Desired Logic. They will need prompting to get started, so you might want to show some of the sample goals in Figure 8.6, as an example.

Beware of using a "canned" list of all possible goals. If you do, your group will stop any individual thinking and just copy goals from your list. Use Figure 8.6 (or excerpts from it) only as a guide to illustrate that a goal:

Is a major element of work

Is a complete sentence

Can be understood when read stand alone

When each of your group has a list of goals that he or she has defined, have them copy the list onto a flip chart. Then as a group, review all the goals (on flip charts). Cross out those that are redundant. Clean up the wording and meaning of each goal that is retained.

At this point you should have at least 40 and probably not more than 90 goals. Leave your goal charts on the wall, but take down everything else (except the study agenda) and send it in for typing.

ESTABLISHING THE CRITERIA FOR THE SELECTION OF TASKS

In this section you will sequence the goals by dependency relationships, estimate goal durations, calculate goal start–stop dates, and identify your time fence for task selection.

Your goals are not necessarily numbered in the sequence that they should be accomplished, so your first step is to get them in the right sequential relationship. You will end up with a PERT chart, with one variation. All goal lines are to be horizontal, because it is easier for the typist (who prepares your final report) to type in the goal number.

Assume that in the previous section you identified 35 goals. It really does not matter which goal you select first to sequence the rest of the goals. Let's say that you pick goal number 9 as a start. Draw a horizontal line and number it "9" as shown by item "A" in Figure 8.7. Then pick any other goal, say 15, and ask your group whether goal 15:

Has to be done before goal 9? (See B, Figure 8.7.)

Has to be done after goal 9? (See C, Figure 8.7.)

Has to be done concurrently with goal 9? Note that a yes on this choice has other implications. Do the start nodes match? Do both start and finish nodes match? Do no nodes match? See D, Figure 8.7.

Continue on with the next goal, asking the same questions. Do not worry about the neatness of your first draft; it is bound to be messy. Later you will make a clean, neat version, but right now you just want to get the relationships down on paper.

Let's consider what is meant by "the right dependency relationships." There are two purist definitions:

A logical relationship of goals reflecting what *must be done* before what else, as opposed to what we'd *like to do* before what else.

A logical relationship of goals reflecting what we think we can do before what else.

Neither of the above two definitions is correct. You want a realistic balance between these two extremes. Tell your group that the first viewpoint is the basic ground rule, but to temper it with reality. For example, let's say that goals 9 and 15 should be done concurrently, because they both have to be complete (common ending node) before goal 22 can start. But your group tells you that it is a known fact that the only person who can do goals 9, 15, and 22 is Leslie Bastile. Well, if that is truly a fact (that's the key—is that really a fact?), then let's not fool ourselves. Sequence the goals as to how they would be done—9, then 15, and finally 22—all in one straight line. (See E, Figure 8.7.)

You may have noticed that we have not, at this point, estimated the durations of the goals. NEVER estimate durations prior to sequencing elements of work. If you do, you will find your group saying, "Oh, let's see, goal 9 has the same duration as goal 22, so why don't we show them as being done in parallel." You are trying to depict a realistic dependency, not dependency for convenience. The implementation plan will assure that work is loaded to personnel in a practical way.

Figure 8.8 is a sample goal network. Note that there are some dashed arrows on this chart. They are used to illustrate a controlled dependency. For example, in Figure 8.8, goal 25 is dependent on goals 24, 33, and 4. Goal 15 is dependent on goals 17, 26 and 4.

The next thing that you are going to do is estimate your goal durations. temporarily take down the flip chart of your network and put it away. You want your duration estimates to reasonably reflect the durations for your goals. If you leave the network in full view, what will happen is this. Look at Figure 8.8. Let's say that your group estimated two months for goal 11. When they get to goal 27, guess what the duration estimate will be? The thinking is this:

1. If goal 27 is really less than two months, let's assign two months to it anyway, since we have the time (constrained by goal 11).
2. If goal 27 is really more than two months, let's assign two months to it. We'll eliminate any slack (or float) in goal 11 and shorten the total project duration.

For this reason, take down the network when you estimate goal durations.

To estimate goal durations, make up a flip chart with the following headings.

		Estimated Dates	
Goal Number	Estimated Duration in Weeks	Start	Stop

Enter the number of each goal under "Goal Number." Your goal flip charts are still on the wall. Have your group agree (100% agreement) on the number of weeks that it would take to accomplish each goal. This is not *man* days or *man* weeks. It is a rough estimate of the total number of weeks that it would take for *this company* to accomplish the goal. There is no doubt that your group will say, "Let's see, to do this goal, Leslie, Jack, Bill, and Sally will be working on it, so it should take three weeks." That's okay. You want to be as realistic as you can with this rough cut, gut feel.

Be sure to put this disclaimer on your chart of goal durations: "These estimates do not reflect actual plan dates. They are only a preliminary step to obtaining plan dates." If you do not do this, someone may walk into the study room, see your dates, and then spread the word that the "XX" System will be implemented by XX/XX/XX. That is obviously not right. The whole reason for this exercise is to figure out which goals are going to be subdivided into tasks. The only time that goal or task plan dates become valid is after the implementation planning phase.

At this point, you have estimated your goal durations in weeks and you have a flip chart that looks like Figure 8.9.

You'll note that I told a small lie earlier. Under the definition of a goal, I stated that the minimum duration was a month and now I am saying that you should do the estimate in weeks. Well, that was done to establish thinking in terms of "months of work." But the durations are going to vary. Again, you want to be practical, so if the stated goal can have a duration of three weeks, there is no point in force fitting the duration to a one-month minimum.

Put your goal network back up on the wall and add the durations to the network. The start date for the total study is zero. You should calculate dates in relative time, instead of actual dates. That way the actual date on which the implementation starts does not matter, because the goal offset dates will still apply. Every goal with a start that touches the start node is assigned the zero start date.

For example, if you were using the network in Figure 8.8, goals 14, 2, 30, 3, 12, 23, and 19 would have a start date of zero. Your estimated durations chart would look like Figure 8.10.

For every goal that has this zero start date, add the weeks of duration and write in the stop date as in Figure 8.11.

Goal 35 cannot be started until goal 14 is complete (Figure 8.8 sample goal network). If the stop date for goal 14 is 8, then the start date on goal 35 is 8. Likewise:

The Start of	Depends on the Stop of
28	2
31	30
7	3
10	3
22	3
26	3
13	12
18	23
21	23

After extending these start dates by durations, your chart would look like Figure 8.12.

Your completed charts should look like Figures 8.13 and 8.14.

It is obvious from the example used, that the longest path through the network is 56 weeks. From the back of the network to the front, this longest path consists of goals 6, 8, 29, 13, and 12. This is called the "critical path." If any of these goals incur delays in starting, or require additional time, the total project will slip past the scheduled completion date.

The critical path can also be calculated. Start by determining the largest number in the estimated stop date column. The total project cannot be completed prior to this date. Enter the largest early stop date as the latest stop date of all goals that directly contact the end node (Figure 8.14). As shown in Figure 8.15, the date of 56 is entered as the stop date for nine goals. Subtract the goal duration from the late stop date to obtain the late start date as shown in Figure 8.16.

The late start date of goal 1 is the late stop date of goal 20. Figure 8.17 illustrates the logic for bringing the late dates through one more set of goals.

Figure 8.18 shows the completed calculation. Float (or slack) is the difference between the early stop date and the late stop date. All goals with a float of zero are critical goals. If they are deferred for any reason, the entire project will slip.

Now that you know the sequence relationship of the goals, the estimated goal durations, and the related goal start–stop dates, you can identify which goals should be divided into tasks.

Your group will have to select a time fence with which they feel comfortable. Normally on a first study, this time fence is six to nine months from the anticipated implementation start date. This is not a hard rule. The extremes are three to 18 months. Recognize, however, that the Implementation Plan must be revised when either of the following occurs:

Due to slippages, invalid estimates, and/or personnel changes, the original plan is no longer valid.

The identified tasks have been accomplished (the end of the time fence has been reached), and now more goals must be subdivided into tasks.

Let's start by assuming that your group agrees on a time fence of six months (26 weeks). The goals that fall within this time frame (by late stop date) are as follows:

Goal	Late Start	Late Stop	Float
2	14	24	14
3	14	17	14
12	0	4	0
13	4	8	0
14	17	25	17
22	17	20	14
29	8	14	0

Seven out of 35 goals are selected for subdividing into tasks. The ratio of 7 out of 35 is acceptable, but your group may want to do more or less work. If they desire to subdivide more than the 7 out of 35 ratio, then raise the 26-week time fence to 28, 30, or whatever it takes to get a realistic comfort level. If they desire to subdivide less than the selected goals, then check the float that you have for each goal. In the example, goal 14 has the highest amount of float (17 weeks). It should be backed out first.

At this point, you are tuning at a rough level, to obtain a comfortable work load for your group. If you select 20 to 30% of your identified goals to be subdivided into tasks, then you are in the proper range.

You now know which goals are going to be used to build your front-end detailed implementation plan of tasks. Keep your goal flip charts and your goal network chart on the wall. All other charts can be submitted to typing.

FIGURES

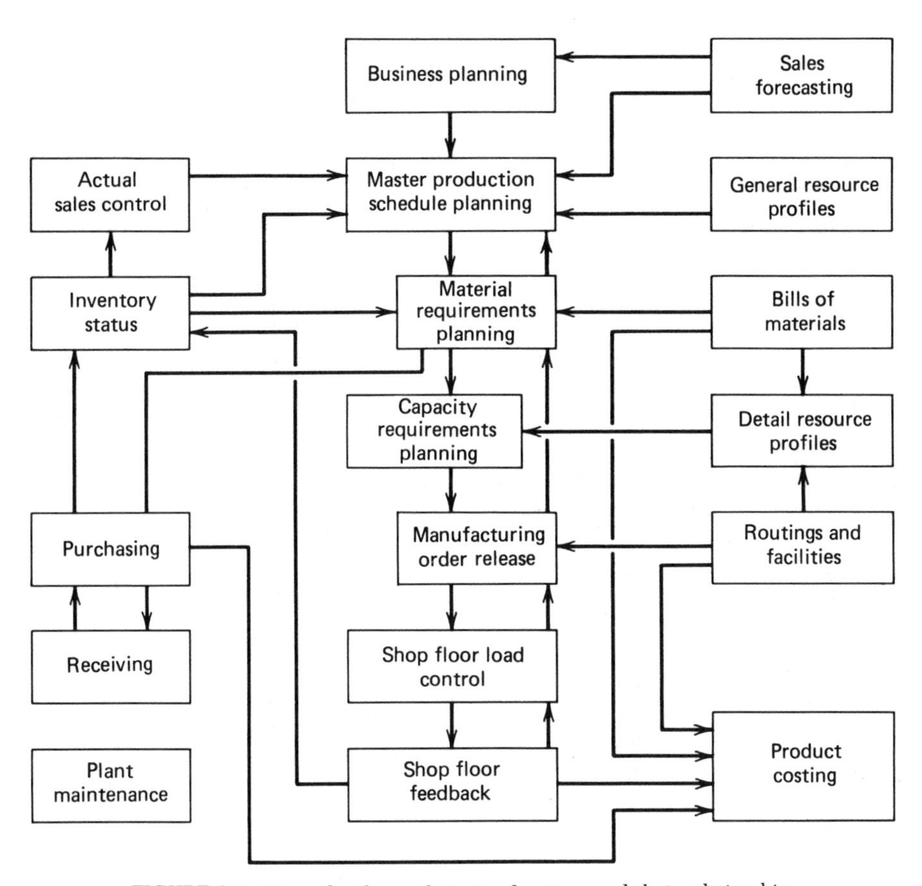

FIGURE 8.1 Example of manufacturing functions and their relationship.

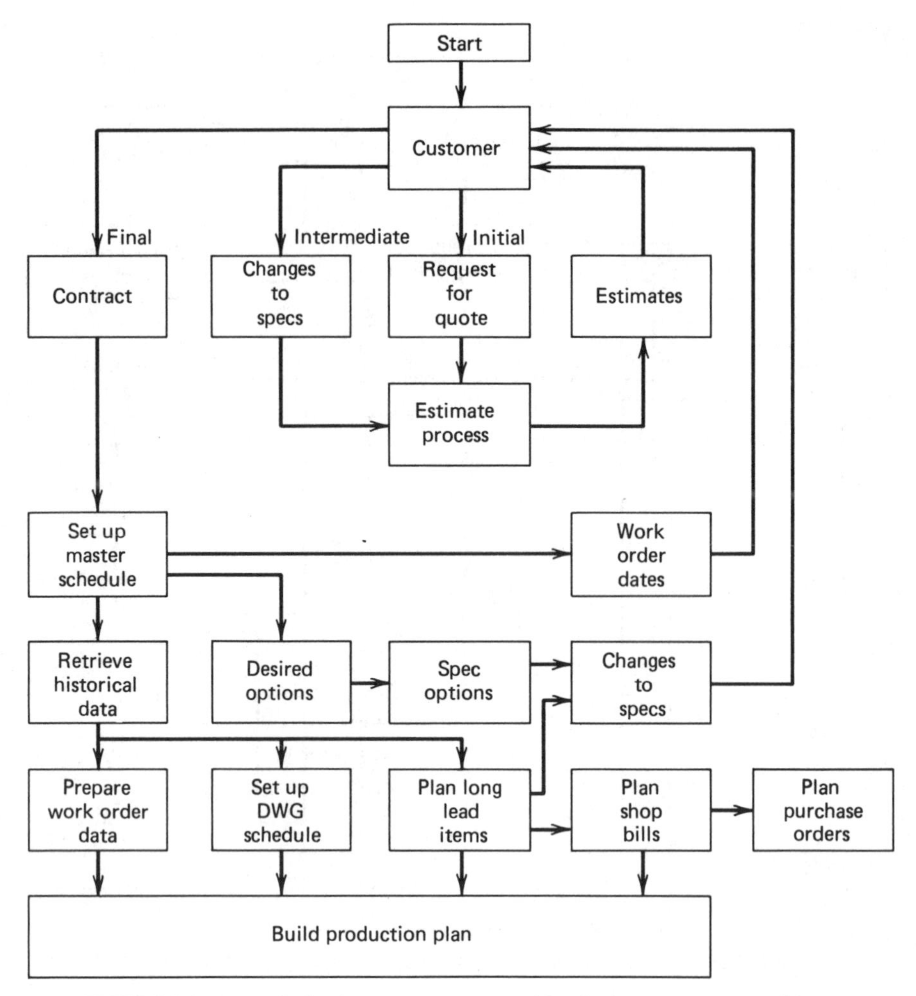

FIGURE 8.2 Example for documenting a general level of current operational logic.

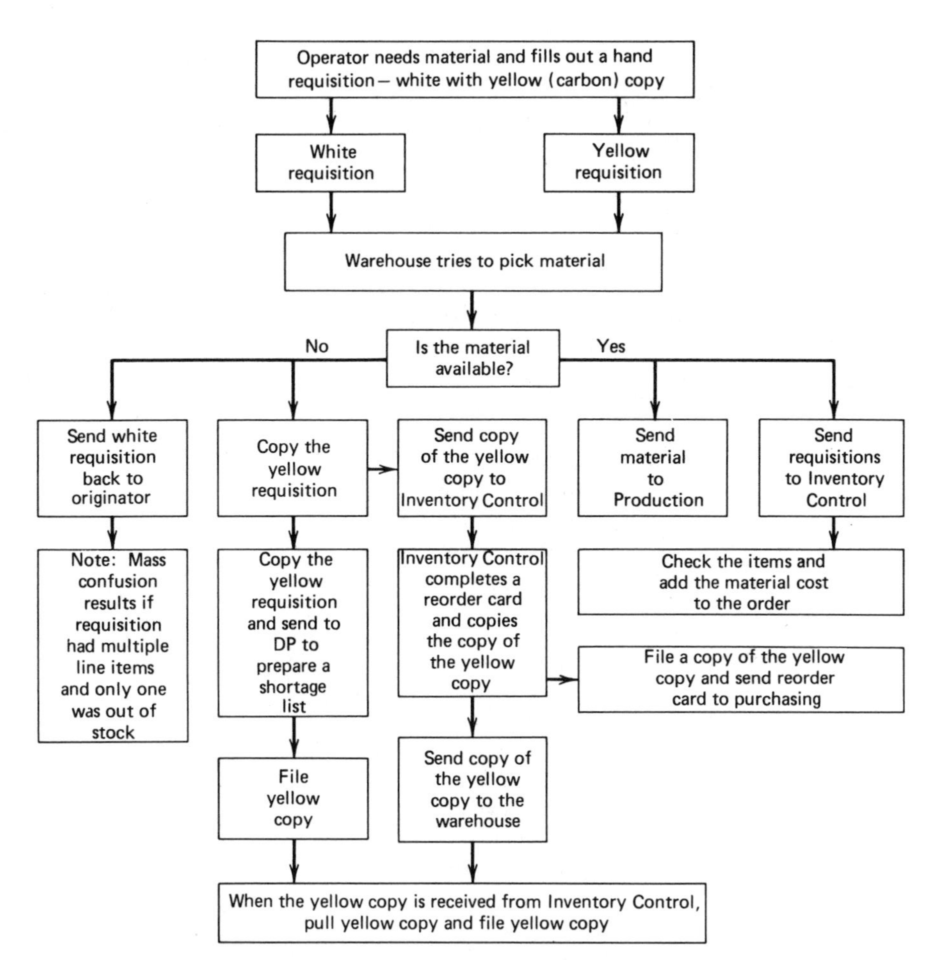

FIGURE 8.3 An expansion of obtain material from warehouse.

Drawings
Shop bills
Work orders
Steel bills
Material requisitions—white
Material requisitions—yellow
Material requisitions—green
Material requisitions—pink
Material requisitions—blue
Recut orders—white
Recut orders—pink
Shop bill order forms
Purchase orders
Computer runs (20 to 30 reports)
Service department requisitions
Tool room requisitions
Employee authorization forms
Inspector's reports of individual Welder
Welders I.D. forms
Quality control forms
Engineering change requests
Time cards
Dispensary passes
Employee warning notices
Call in slips
Transfer of yard personnel
Employee absent from work forms
Employee requisition forms
Waivers to pay holiday pay
Performance appraisals
Authorizations to adjust an employee's rate
Training authorizaition forms
Employee separation forms
Requests for early performance review
Safety inspections
Accident report

FIGURE 8.4 Documentation managed by the line foremen.

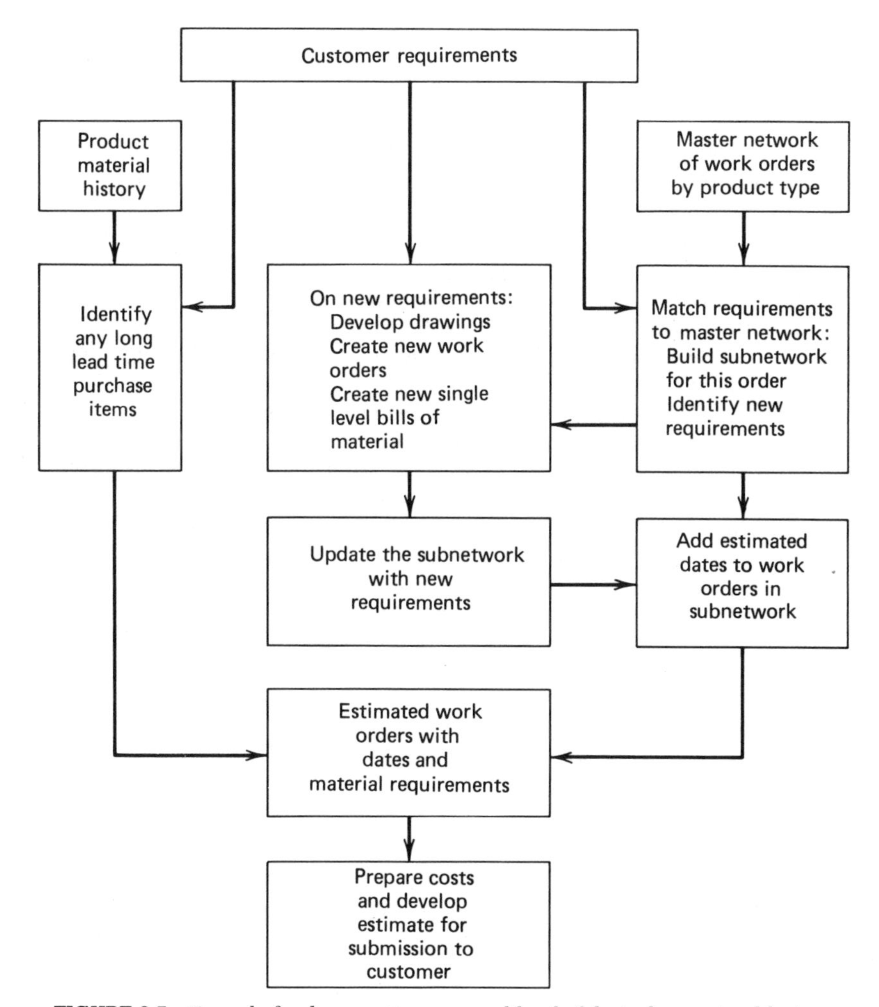

FIGURE 8.5 Example for documenting a general level of desired operational logic.

Identify and establish the physical constraints necessary to secure the warehouse.
Define and implement the procedures to maintain controls on who has access to material in the warehouse.
Develop communication procedures between production, engineering, material control, production planning, and general planning to easily and effectively communicate changes.
Install the selected inventory accounting software, test the software and train personnel on the sample data base.
Install warehouse terminals to handle receipts, issues and adjustments. Train personnel on the use of the terminals and the software.
Develop procedures to allow for all departments to be involved in planning proposed new construction work.
Define the criteria, procedures, and time parameters to be used for cycle counting.
Reorganize the DP organization to provide more specialization and separation of duties.
Define the procedures to measure inventory levels, so that they can be checkpointed at any point in time. This is to include a dynamic recalculation of safety stock based on demand fluctuations.
Pilot test inventory accounting in the small parts warehouse.
Increase the size and quality of the DP staff through hiring and technical training.

FIGURE 8.6 Sample goals.

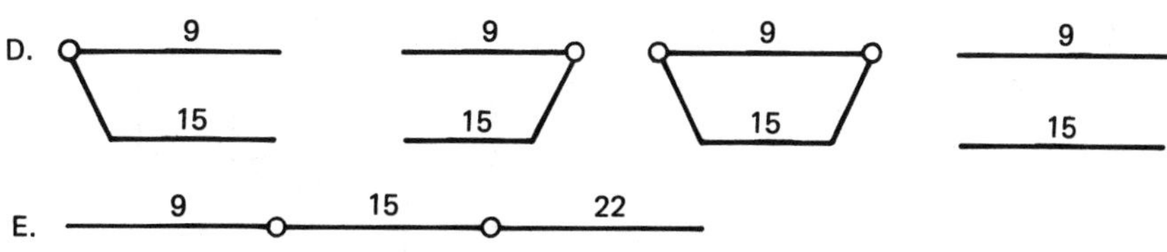

FIGURE 8.7 Sequencing goals.

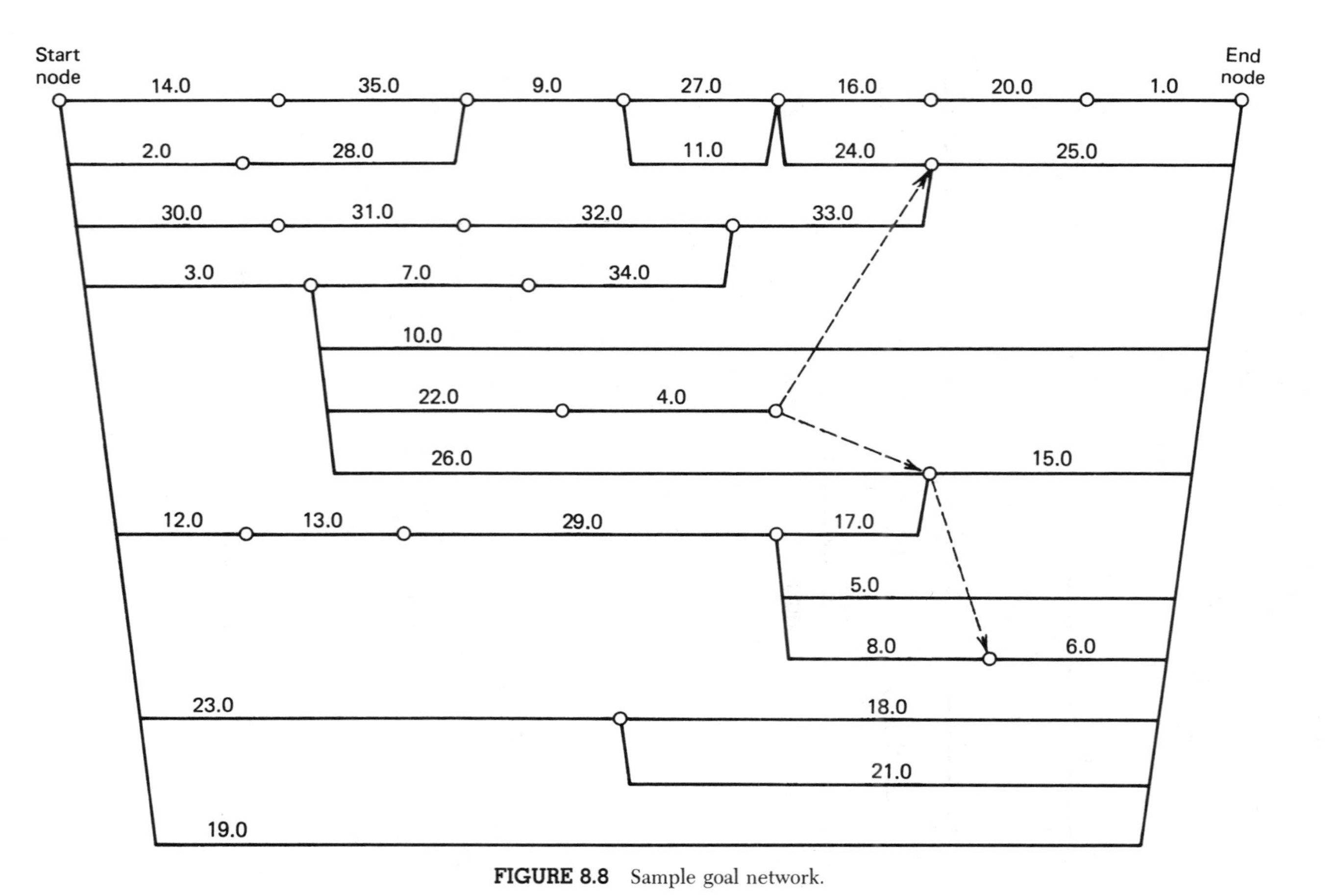

FIGURE 8.8 Sample goal network.

Goal Number	Estimated Duration in Months	Estimated Dates		Goal Number	Estimated Duration in Months	Estimated Dates	
		Start	Stop			Start	Stop
1	4			21	4		
2	10			22	3		
3	3			23	5		
4	12			24	6		
5	16			25	8		
6	24			26	7		
7	6			27	12		
8	18			28	4		
9	2			29	6		
10	9			30	8		
11	6			31	4		
12	4			32	4		
13	4			33	8		
14	8			34	9		
15	9			35	3		
16	3						
17	12						
18	2						
19	5						
20	4						

FIGURE 8.9 Estimation goal dates. (These estimates do not reflect actual plan dates. They are only a preliminary step to obtain plan dates.)

Goal Number	Estimated Duration in Months	Estimated Dates		Goal Number	Estimated Duration in Months	Estimated Dates	
		Start	Stop			Start	Stop
1	4			21	4		
2	10	0		22	3		
3	3	0		23	5	0	
4	12			24	6		
5	16			25	8		
6	24			26	7		
7	6			27	12		
8	18			28	4		
9	2			29	6		
10	9			30	8	0	
11	6			31	4		
12	4	0		32	4		
13	4			33	8		
14	8	0		34	9		
15	9			35	3		
16	3						
17	12						
18	2						
19	5	0					
20	4						

FIGURE 8.10 Estimating goal dates (continued). Note that these estimates do not reflect actual dates; they are only a preliminary step to obtain plan dates. The dates are earliest possible dates.

Goal Number	Estimated Duration in Months	Estimated Dates		Goal Number	Estimated Duration in Months	Estimated Dates	
		Start	Stop			Start	Stop
1	4			21	4		
2	10	0	10	22	3		
3	3	0	3	23	5	0	5
4	12			24	6		
5	16			25	8		
6	24			26	7		
7	6			27	12		
8	18			28	4		
9	2			29	6		
10	9			30	8	0	8
11	6			31	4		
12	4	0	4	32	4		
13	4			33	8		
14	8	0	8	34	9		
15	9			35	3		
16	3						
17	12						
18	2						
19	5	0	5				
20	4						

FIGURE 8.11 Estimating goal dates (continued).

Goal Number	Estimated Duration in Months	Estimated Dates		Goal Number	Estimated Duration in Months	Estimated Dates	
		Start	Stop			Start	Stop
1	4			21	4	5	9
2	10	0	10	22	3	3	6
3	3	0	3	23	5	0	5
4	12			24	6		
5	16			25	8		
6	24			26	7	3	10
7	6	3	9	27	12		
8	18			28	4	10	14
9	2			29	6		
10	9	3	12	30	8	0	8
11	6			31	4	8	12
12	4	0	4	32	4		
13	4	4	8	33	8		
14	8	0	8	34	9		
15	9			35	3	8	11
16	3						
17	12						
18	2	5	7				
19	5	0	5				
20	4						

FIGURE 8.12 Estimating goal dates (continued).

Goal Number	Estimated Duration in Months	Estimated Dates		Goal Number	Estimated Duration in Months	Estimated Dates	
		Start	Stop			Start	Stop
1	4	35	39	21	4	5	9
2	10	0	10	22	3	3	6
3	3	0	3	23	5	0	5
4	12	6	18	24	6	28	34
5	16	14	30	25	8	34	42
6	24	32	56	26	7	3	10
7	6	3	9	27	12	16	28
8	18	14	32	28	4	10	14
9	2	14	16	29	6	8	14
10	9	3	12	30	8	0	8
11	6	16	22	31	4	8	12
12	4	0	4	32	4	12	16
13	4	4	8	33	8	16	24
14	8	0	8	34	9	9	18
15	9	26	35	35	3	8	11
16	3	28	31				
17	12	14	26				
18	2	5	7				
19	5	0	5				
20	4	31	35				

FIGURE 8.13 Estimating goal dates (continued).

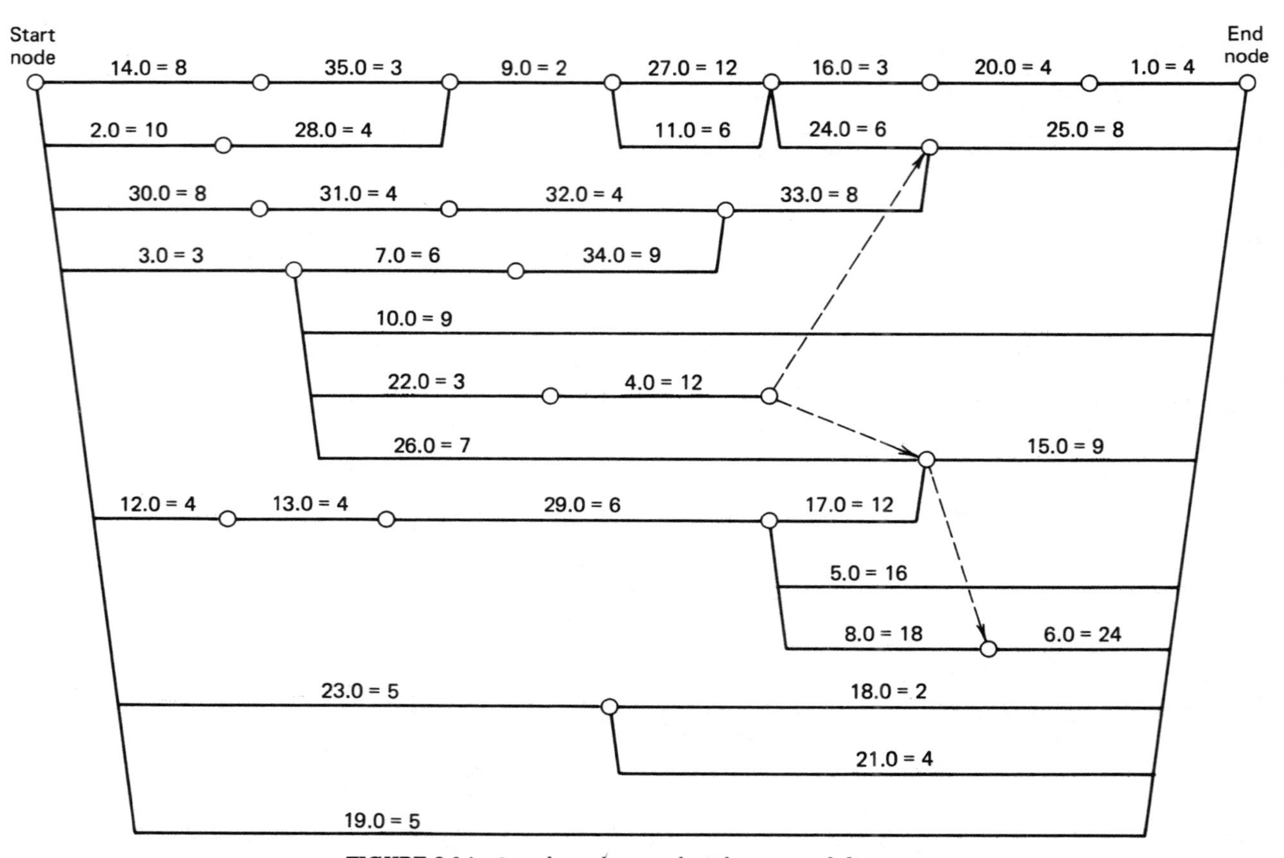

FIGURE 8.14 Sample goal network with associated durations.

Goal	Duration	Early Start	Early Stop	Float	Late Stop	Late Start
1	4	35	39		56	
2	10	0	10			
3	3	0	3			
4	12	6	18			
5	16	14	30		56	
6	24	32	56		56	
7	6	3	9			
8	18	14	32			
9	2	14	16			
10	9	3	12		56	
11	6	16	22			
12	4	0	4			
13	4	4	8			
14	8	0	8			
15	9	26	35		56	
16	3	28	31			
17	12	14	26			
18	2	5	7		56	
19	5	0	5		56	
20	4	31	35			
21	4	5	9		56	
22	3	3	6			
23	5	0	5			
24	6	28	34			
25	8	34	42		56	
26	7	3	10			
27	12	16	28			
28	4	10	14			
29	6	8	14			
30	8	0	8			
31	4	8	12			
32	4	12	16			
33	8	16	24			
34	9	9	18			
35	3	8	11			

FIGURE 8.15 Calculating the critical goals.

		Early			Late	
Goal	Duration	Start	Stop	Float	Stop	Start
1	4	35	39		56	52
2	10	0	10			
3	3	0	3			
4	12	6	18			
5	16	14	30		56	40
6	24	32	56		56	32
7	6	3	9			
8	18	14	32			
9	2	14	16			
10	9	3	12		56	47
11	6	16	22			
12	4	0	4			
13	4	4	8			
14	8	0	8			
15	9	26	35		56	47
16	3	28	31			
17	12	14	26			
18	2	5	7		56	54
19	5	0	5		56	51
20	4	31	35			
21	4	5	9		56	52
22	3	3	6			
23	5	0	5			
24	6	28	34			
25	8	34	42		56	48
26	7	3	10			
27	12	16	28			
28	4	10	14			
29	6	8	14			
30	8	0	8			
31	4	8	12			
32	4	12	16			
33	8	16	24			
34	9	9	18			
35	3	8	11			

FIGURE 8.16 Calculating the critical goals (continued).

		Early			Late	
Goal	Duration	Start	Stop	Float	Stop	Start
1	4	35	39		56	52
2	10	0	10			
3	3	0	3			
4	12	6	18		32	
5	16	14	30		56	40
6	24	32	56		56	32
7	6	3	9			
8	18	14	32		32	
9	2	14	16			
10	9	3	12		56	47
11	6	16	22			
12	4	0	4			
13	4	4	8			
14	8	0	8			
15	9	26	35		56	47
16	3	28	31			
17	12	14	26		32	
18	2	5	7		56	54
19	5	0	5		56	51
20	4	31	35		52	
21	4	5	9		56	52
22	3	3	6			
23	5	0	5		52	
24	6	28	34		48	
25	8	34	42		56	48
26	7	3	10		32	
27	12	16	28			
28	4	10	14			
29	6	8	14			
30	8	0	8			
31	4	8	12			
32	4	12	16			
33	8	16	24		48	
34	9	9	18			
35	3	8	11			

FIGURE 8.17 Calculating the critical goals (continued).

Goal	Duration	Early Start	Early Stop	Float	Late Stop	Late Start	Critical Goals
1	4	35	39	17	56	52	
2	10	0	10	14	24	14	
3	3	0	3	14	17	14	
4	12	6	18	14	32	20	
5	16	14	30	26	56	40	
6	24	32	56	0	56	32	*
7	6	3	9	21	31	25	
8	18	14	32	0	32	14	*
9	2	14	16	14	30	28	
10	9	3	12	34	56	47	
11	6	16	22	20	42	36	
12	4	0	4	0	4	0	*
13	4	4	8	0	8	4	*
14	8	0	8	17	25	17	
15	9	26	35	21	56	47	
16	3	28	31	17	48	45	
17	12	14	26	8	32	20	
18	2	5	7	49	56	54	
19	5	0	5	51	56	51	
20	4	31	35	17	52	48	
21	4	5	9	47	56	52	
22	3	3	6	14	20	17	
23	5	0	5	47	52	47	
24	6	28	34	14	48	42	
25	8	34	42	14	56	48	
26	7	3	10	22	32	25	
27	12	16	28	14	42	30	
28	4	10	14	14	28	24	
29	6	8	14	0	14	8	*
30	8	0	8	24	32	24	
31	4	8	12	24	36	32	
32	4	12	16	24	40	36	
33	8	16	24	24	48	40	
34	9	9	18	22	40	31	
35	3	8	11	17	28	25	

FIGURE 8.18 Calculating the critical goals (continued).

CHAPTER NINE

PHASE III: IMPLEMENTATION PLANNING

PURPOSE/OBJECTIVES OF IMPLEMENTATION PLANNING

The successful completion of this phase is a key element in the success of your study effort. All too many companies have grand plans for the future. They will show you a fantastic, integrated, distributed, on-line information system plan. It will exist in five years. Or will it? So far all that they have is the design. They know where they want to be in five years, but they don't know what they have to do tomorrow to get there.

The purpose of the implementation planning phase is to build a total plan that is detailed in the front end (describing what to do tomorrow) and more general the further out in time that the plan goes.

To successfully develop a plan, you will have to do the following:

1. Subdivide selected goals into defined tasks.
2. Modify the goal dependency relationship chart (network) to reflect tasks as well as goals not subdivided into tasks. This will allow you to maintain the "total picture" of the work to be performed.
3. Assign performance responsibility to specific individuals and estimate task durations.

4. Next, you load level tasks by person and by department. This is the step that puts reality into your plan as opposed to listing 50 things that a person has to do sometime during the next quarter.
5. Finally, document your plan so that it can be understood by both the "doer" and by the "reviewer."

HOW TO SUBDIVIDE GOALS INTO TASKS

The purpose of this section is to identify those finite elements of work (tasks) that can be performed in the immediate future.

The most effective way to subdivide the identified goals into tasks is to have it done individually by members of your study group. List the goal numbers to be subdivided on a flip chart and ask for volunteers. You will find that a person will volunteer for the goal(s) about which he or she has the most knowledge. This is exactly what you want. Now give your team the following guidelines:

A task is an element of work necessary to achieve the goal.

The duration of a task is one to ten days. This is duration, not mandays. For example, here is task A. It looks like Bill should be responsible for doing this one. Will Bill actually do it himself? No, he'll probably assign Jack and Sue to do it. How long will it take, if they both work on it? Well, it will probably take nine days. (Note that this is a "best estimate" of duration at this time and for general sizing purposes only.) Okay, so task A, with its description and duration is a valid task, even though it takes 18 mandays.

A task description should be a complete sentence. It should be easily understood when read by itself. For example, let's say that we have these two tasks:

A. Recalculate the lot sizes for items in class X and Z.

B. Enter the above into the Requirements Planning System.

Now, let's say Kevin performs task A. It is in his task "To Do" list. Task B is going to be done by Mary. It is in her "To Do" list. There is no way that she can understand what "the above" means in task B unless she goes back to the total final study report. Task B does not stand alone.

Tasks do not have to be written (numbered or sequenced) in any specific order. Dependency relationships will be established at a later point in the study. Therefore, write down the tasks as they are thought of.

Although the assignment of who is going to be responsible for making sure that a task is done is not going to be assigned now, the study group member writing the task should have a fair idea of who that responsible person will be.

Since tasks have an estimated duration, they have an end. This end should be one that can be clearly identified. A task that is stated "Monitor the interplant movement of materials," has no value. When does it ? How will you ever be able to measure performance to plan if tasks never end or the end is vague? A better way to state the above task is to say "Implement a procedure for the forklift and truck dispatcher to maintain a log for seven days consisting of:

Time move requested
Move from location
Time move started
Time move completed
Move to location

Whenever the above task is scheduled, seven days later it can be determined as to whether or not it is complete.

You should make up a temporary flip chart on task guidelines, such as the one in Figure 9.1, and leave it on the wall (after you've gone through the above explanation) as a reference while your team is writing tasks.

Remember that although the tasks do not have to be sequenced in any specific order (when they are being written), they should relate back to the original goal that they support. For example, if goal number 6 had been, "Install an Inventory Accounting System," then all of the tasks identified to support that goal should be numbered 6.1, 6.2, 6.3, 6.4, and so on, as illustrated in Figure 9.2.

This will allow you to look at any task in the future and relate it back to the goal for which it was originally generated.

When each member of your group completes writing all of the tasks for his assigned goals, have them copied on flip chart paper. As a total group, review all the created tasks. Check for the following:

1. Is the meaning of the task clear?
2. Does the task meet the original task criteria (refer to Figure 9.1)?
3. Is this task redundant to another and should therefore be eliminated?
4. Are there any other tasks that should be written to support this goal?

Frequently, when you go through this process, you will identify stand-alone tasks. These are tasks that do not quite fit in with the goal being defined. In fact, they either do not fit with any defined goal, or they fit with multiple goals. An interface task is an example. In this situation, you have two choices. You can lump all tasks of this type in a general category, or you can identify the task under the goal to which it applies the most. It really does not matter which approach is chosen, as long as the task is documented. Later on, all of the tasks will be sequenced according to when they should be done and by whom.

At this time, hanging on the walls of your study room should be the goal flip charts, the goal network chart, and all of your task description charts. Any other material, like the critical path calculation charts, can be sent to typing.

BUILDING A TASK/GOAL DEPENDENCY RELATIONSHIP CHART

In this step, you sequence the tasks according to their dependency relationships, using the same rules that you applied when you sequenced goals. The sample goal network is shown in Figure 9.3 for your review.

In the previous section, you identified those goals that you wanted to subdivide into tasks to obtain a detailed front-end plan. Circle those goals on the goal network, as shown in Figure 9.4.

Now for an additional guideline. In the network example, goal 12 directly precedes goal 13. When you did the goal network, this was your most convenient division. But now that these goals are being subdivided into tasks, that rigid division line need no longer exist. Goal 12 could have a task that overlaps goal 13, or even extends to the end of the network.

A simple way to perform this step is to initially review the tasks per goal individually. For example, take your list of tasks for goal 12 and sequence those as they relate to each other. Do the same for the tasks on all other selected goals. Next, see if one set of tasks, such as for goal 12, fits into the goal network where goal 12 was located before it was subdivided. When you complete this process, you will have one network that reflects dependency relationships of all tasks, plus all goals not subdivided into tasks as shown in Figure 9.5.

Note that in Figure 9.5, an element of work called 12.5 is task 5 of goal 12. Also, you will notice that no durations are shown. You have not yet assigned durations to the tasks. The goal durations have been deleted so that a pure dependency relationship can be maintained, and you will want to reevaluate your original duration estimates for goals.

Temporarily take your network down from the wall and put it out of sight. You will next be estimating durations, and you do not want the dependency relationships to have an impact on estimated durations.

HOW TO ASSIGN TASKS TO INDIVIDUALS AND ESTIMATE TASK DURATIONS

Your purpose in this section is to derive duration estimates that are as realistic as possible. Remember when I established the definition of a task? I said that it could have a duration of one to ten days. Well, the one- to ten-day constraint was imposed as a learning tool, to set the thinking pattern to that time frame. Now you want to obtain estimates that are as realistic as possible—12 days, 20 days, or even 30 days. Caution: If you get an estimated task duration of over 30 days, the task should probably be subdivided into smaller segments of work.

Set up a flip chart page using the headings shown in Figure 9.6 and number in your identified goals and tasks. Of the 9 goals shown in Figure 9.6, notice that goals 2 and 3 are listed as the subdivided tasks. The column heading used on Figure 9.6 is multipurpose, not all of which will be completed in this section.

You still have your goal and task descriptions on the wall. (Your network is not on the wall.) Have your group identify the person best qualified to be responsible for making sure that each element of work (task) gets performed. When you are finished, your flip chart should look like Figure 9.7.

The best way to get a realistic duration estimate is to ask the person who will probably be doing the work. For example (referring to Figure 9.7), call in Mr. K. Brown and ask:

1. Do you agree that you should be responsible for accomplishing tasks 2.1, 2.6, and 3.2? (Given a yes on this question, you continue.)
2. How long, in days of duration, do you think that it will take you to get this element of work done?
3. During the estimated duration, what percentage of your time do you think that you will be working on the task?

Try to restrict your duration estimates to a minimum of one day with a minimum of 50% of time applied to the task. If you allow less than the above, you will get hung up in calculating a lot of fractions. As you interview each "assigned person," add the days of duration and percentage of time to your flip chart. Also, calculate the mandays of labor (days of duration times percentage of time), and add that to your chart, even though you will not be using the manday column until later. When you have completed the estimates for tasks, your charts would look like those in Figure 9.8.

There is something else that you might want to consider. The examples that have been illustrated have shown an assigned person on goals. You may not always be able to do this. Sometimes, the best that you can do on a goal is to identify the department that will be responsible. However, if you did identify people generally responsible for goals (knowing that the goals will later be divided into tasks), while you are doing the task duration interviews, get those people to have your original goal durations reestimated. If you can do this, then you will have more accurate labor data for your project cost/benefit analysis. Your chart now looks like Figure 9.9.

Note that in Figure 9.9, all time has been converted to days. Although this gives you some large numbers (goal durations) that you relate to some small numbers (task durations), the arithmetic is easier if your entire base is in days.

At this time, update Figure 9.5 by adding the durations for all tasks and goals. The sample task/goal network is shown in Figure 9.10.

Leave all flip charts on the wall. This includes:

Goal descriptions
Task descriptions
Task/goal, assigned person, and duration charts
The task goal network

HOW TO LOAD LEVEL TASKS BY PERSON WITHIN DEPARTMENTS

This is an extremely important step. It injects reality into your implementation plan. It also creates a very positive attitude within your study group—that this whole project really can be accomplished.

The purpose of load leveling is to spread the tasks and goals by person without violating the network dependency relationships, or overloading any one person. You should recognize that everything you've done so far in this phase has been done to get to this point. All material prepared to date is for *guideline purposes* only. At this point, anything can change. Durations can be adjusted. Percentages of time devoted to a task can be adjusted. Tasks can be split up or combined. You want to be flexible going into this phase to make the output as realistic and implementable as possible.

You, or your group, should know going into this step approximately how many mandays are available for consumption by the project. If you do not know this "accepted amount," then you will probably make at least two passes at load leveling. You'll do your first pass and produce what you think is a very viable plan. Your executive sponsor may make one or more of these valid points:

1. It looks like department XX is a bottleneck. I can let you have twice the amount of labor from that department than you used for your calculations. Tell me how much that will shorten the total implementation.
2. The total implementation is going to take too long. Show me where I can best apply additional labor (and how much) so that we can speed up the implementation.
3. I can't afford to apply the resources that you indicated from department YY. Cut that amount by 50% and tell me what the impact is on the total implementation date.

If you know, going into the load-leveling step, how much labor is likely to be available, you'll usually be able to do this step in one pass. If not, you are almost guaranteed at least two passes.

Occasionally you will find an executive sponsor who likes to play "What If." What if I add people here? (Redo the plan.) What if I reduce people there? (Redo the plan.) What if I don't spend the education money until six months later? (Redo the plan.) You should explain very diplomatically that the most you can do at this stage is to make a maximum of three load-leveling passes. It is very expensive to keep the study group tied up in a manual process of optimizing a load-leveling plan.

Now you are ready to set up for load leveling. Select the longest flat wall without windows that you have in the study room. Mount all of the existing flip charts on the other walls. Paper your blank wall with blank flip chart paper. Use the kind with the light blue one-inch squares on it. One row across the wall is normally enough, as shown in Figure 9.11.

Down the leftmost chart, on the left side, list the personnel that have been assigned responsibilities. Group them by department so that you can accumulate department totals. At the top of the flip charts, starting at the right of the list of names, number the lines of the light blue, one-inch-square grid in even numbers as: 0, 2, 4, 6, 8, 10, and so on. For start-up, your numbers should go up to about 250 or 300. If you need more numbers later on, you can always add more. Your load leveling flip charts should now be structured as shown in Figure 9.12.

Notice in Figure 9.12 that the time scale is in relative time. You have the greatest flexibility and ease of date calculation by just numbering days from 0 to 250. Later, after the plan is completed, you can superimpose actual dates and also take weekends, holidays, and summer plant shutdowns into account.

Now you can actually begin building the plan. To coordinate this effort, you will need people in the following positions:

1. One person working the network. This person is to call out the next task/goal number that can be performed according to dependency relationships, and to check off those completed on the network.
2. One person working the task/goal descriptions. This person is to read out loud the description of the task number called out by the person working the network.
3. One person working the assigned person/duration charts. This person calls out the name of the person to whom the task/goal has been assigned, the duration and the mandays.
4. One person working the load-leveling chart which is being created. This person is to draw a horizontal line representing the duration of the task/goal alongside of the name of the assigned person. Refer to Figure 9.13. J. Smith has already been assigned task 52.5, which ends on day 70. The network person says that task 61.3 cannot start until task 52.5 is complete. Task 61.3 has been assigned to B. Jones. Since 61.3 cannot start until 52.5 is complete, the start of 61.3 is 4 days, so the finish day is 74. The horizontal line is drawn by B. Jones' name starting at day 70 and ending at day 74. The task is then labeled with the task/goal number (61.3), the duration in days (4.0), and the calculated mandays (2.0).
5. One person to act as overall checker. This person is to ensure that the rest of the team is working on the right task/goal for the right assigned person and that it is being plotted in the correct position. This person is important to the coordinated effort, because with any background discussion at all, it is very easy to get out of sync.

All material prepared before load leveling should be used as a guideline for load leveling. Consider these examples to illustrate that point!

1. Tasks 45.1 and 45.2 are to be done in parallel (same start and end points) according to the network. Each one has a duration of 4.0 days and an assigned 4.0 days of labor. They are both assigned to Tammy Robinson.

She obviously cannot work two full-time tasks at the same time. Your two choices are to either reassign one of the tasks to another person or to have Ms. Robinson do the tasks sequentially—45.1 first, followed by 45.2.

2. According to the network, task 52.4 (which is eight days long) must be completed before task 53.8 can start. The tasks have been assigned to different people. Everyone agrees that in reality, all of task 52.4 does not have to be done before task 53.8 can start. In fact if 70 to 80% of 52.4 is finished, then task 53.8 can start. If 52.4 was to start on day 90 and finish on day 98, you should now schedule task 53.8 to start on day 96 instead of day 98.

It cannot be stressed enough that your load leveling should be as realistic as possible. If a person has told you that he or she can apply only 50% of his or her time to the implementation, then you can only load that person up to 50%, even though it may stretch out the total implementation.

When you have completed your load-leveling chart, you should have a reasonably good estimate of when every task/goal will be performed. Now is a good time for a review with your executive sponsor, who will say things like this:

1. It looks like J. Smith is almost always 100% loaded. Would it help if I gave you another person to assist J. Smith?
2. I guess S. Hinkleman told you that he could spend 50% of his time on his project, but from (this date) to (that date), I'm going to have a special assignment for him, so between those dates, reduce his load to 25%.

When you have identified and completed the plan readjustments, your question to the executive sponsor is, "What is a reasonable start date?" Once you have that date, you can overlay the relative time days with actual dates. Take into account weekends, holidays, and any major periods (such as plant shutdowns) when no one is working. You are converting a sequential string of days to a series of dates that reflect actual days (dates) when work will be performed.

By now you realize that your load-leveling chart cannot be typed and included in your final report. Therefore, you need to extract the derived data and convert it to a manageable form. Divide your study group up into three teams to perform these functions.

Team 1

Divide the load-leveling chart up into yearly quarters by drawing vertical lines at March 31, June 30, September 30, and December 31.

Within each quarter, total the mandays per person.

Within each quarter, total the mandays for all personnel within a department to obtain a department total.

Within each quarter, total the department totals to obtain a gross total.

Complete a manday load chart as shown in Figure 9.14.

Team 2

For each assigned person, list the tasks in the sequence that they are to be performed.

When (for one person) multiple tasks have the same start date, list the task with the shortest duration first.

Complete the assigned person task sequence list as shown in Figure 9.15.

Team 3

This team performs the function of overall checker and recorder.

Verify on the assigned person/duration chart (Figure 9.16) that all data are correct as currently reflected on the load-leveling chart.

Using the Assigned person task sequence list developed by Team 2 and the load-leveling chart, complete the planned start and finish date columns on the assigned person/durations chart.

When your three teams are done, you have completed this section. If your load-leveling plan modified some of the original network dependencies, add a note to your network to say, "The final plan may have modified some of the initial identified dependencies."

At this point in your study, you will be faced with a typing bottleneck. To continue with the study, you should have all the data on flip charts also put in a typed format, except the load-leveling chart. If you have a speedy typing service, you may send all the flip charts in for typing. You can also copy the flip charts, so that you can send the originals in for typing and retain the copies in order to proceed with the next section in this phase. Note that the latter choice is normally not a practical alternative. Your last resort is to stop the study until you at least get the rough drafts back from typing.

HOW TO FORMALIZE THE IMPLEMENTATION PLAN

You have done a lot of work and created a lot of detail, but so far you do not have anything that can be "worked from," or "presented from." The individuals who are assigned to perform tasks want to see all the work that they are supposed to do presented in the sequence that it should be done. They do not want the tasks of other people mixed in with theirs, so you will need to prepare individual work plans.

Conversely, the top company executives do not want to review individual work plans. They are interested in when the major elements of the project will start and finish.

Your load-leveling chart is still on the wall. Identify, from your goals, the major elements that will be of interest to the company executives. On the load-leveling chart, mark the beginning and end of each of those major elements. You should restrict this to not less than six items and not more than 12. Prepare a single chart that reflects these key segments of your planned implementation. Refer to the sample in Figure 9.17.

If labor costs are a key concern, you may want to add the mandays per quarter to the bottom of the chart.

You now have a single chart that you can use to explain when each major element of the project is planned to start and finish. This chart is also useful for your study group in the next phase, to determine when costs and benefits are likely to occur in relation to when each element is being implemented.

To prepare the detail implementation pages for every assigned person, you need most of the typed data derived from the load-leveling chart. This can be done by either a "cut-and-paste" method or a "recopy-and-compile" method. Either way, it is a task best done by a person who is not a direct member of your study group, such as your typist.

The steps to prepare the detailed implementation plan are as follows:

Enter the assigned person's name, department, and the date of the last plan update.

Refer to the Assigned Person Task Sequence List (Figure 9.15) and enter the first task for the assigned person.

Refer to the assigned person/duration chart (Figure 9.16) and enter the planned start–stop dates, the duration period, the percentage of time, and the number of mandays.

Refer to the task/goal descriptions and enter the task or goal description.

Select the next task and repeat the process until all tasks/goals are complete for this person.

Select the next person, start a new page, and repeat the above.

See Figure 9.18 as an example of a completed implementation plan page.

At this time, you have accomplished the following:

Identified tasks and goals

Sequenced all tasks and goals in a logical dependency relationship

Assigned personnel to be responsible for the accomplishment of each task and goal

Estimated durations, percentages of applicable time, and mandays

Load-leveled all tasks by person to develop a realistic, achievable plan

Reviewed your plan with your executive sponsor and made any required adjustments

Calculated the required number of mandays for use in cost/benefit planning

Extracted all planning data from the load-leveling chart to build your detail by individual implementation plans

Constructed a project summary chart

Constructed the detailed implementation plans for every assigned person

You have now completed the implementation planning phase. Send all material in for typing except for your project summary chart.

FIGURES

A Task :

Is an element of work
Has a duration of one to ten days
Is a complete sentence
Can be understood when read standing alone
Need not be sequenced with other tasks when it is being written
Should be assignable to one person
Must have a definite ending point

FIGURE 9.1 What a task should be.

Goal No. 6—Install an Inventory Accounting System

Task Number	Task Description
6.1	Have the Supply Foreman and the Inventory Manager attend an inventory control class.
6.2	Hold an in-house training class on stores transaction control.
6.3	Identify the person or group necessary to coordinate and conduct the in-house training on inventory control.
6.4	Establish with management, the criteria for defining and disposing of obsolete parts.
6.5	Define a procedure to dispose of present and future obsolete parts.
6.6	Determine who will be allowed to and responsible for obsoleting parts.
6.7	Design a program, using obsolete parts criteria, to identify present and future obsolete parts.
6.8	Identify the person(s) responsible for reviewing the lot sizing protocol available in the software and define the review periods.
6.9	Obtain a management policy statement on safety stock.
6.10	Assign the person(s) or the group responsible for calculating and managing safety stock levels.

FIGURE 9.2 Sample tasks.

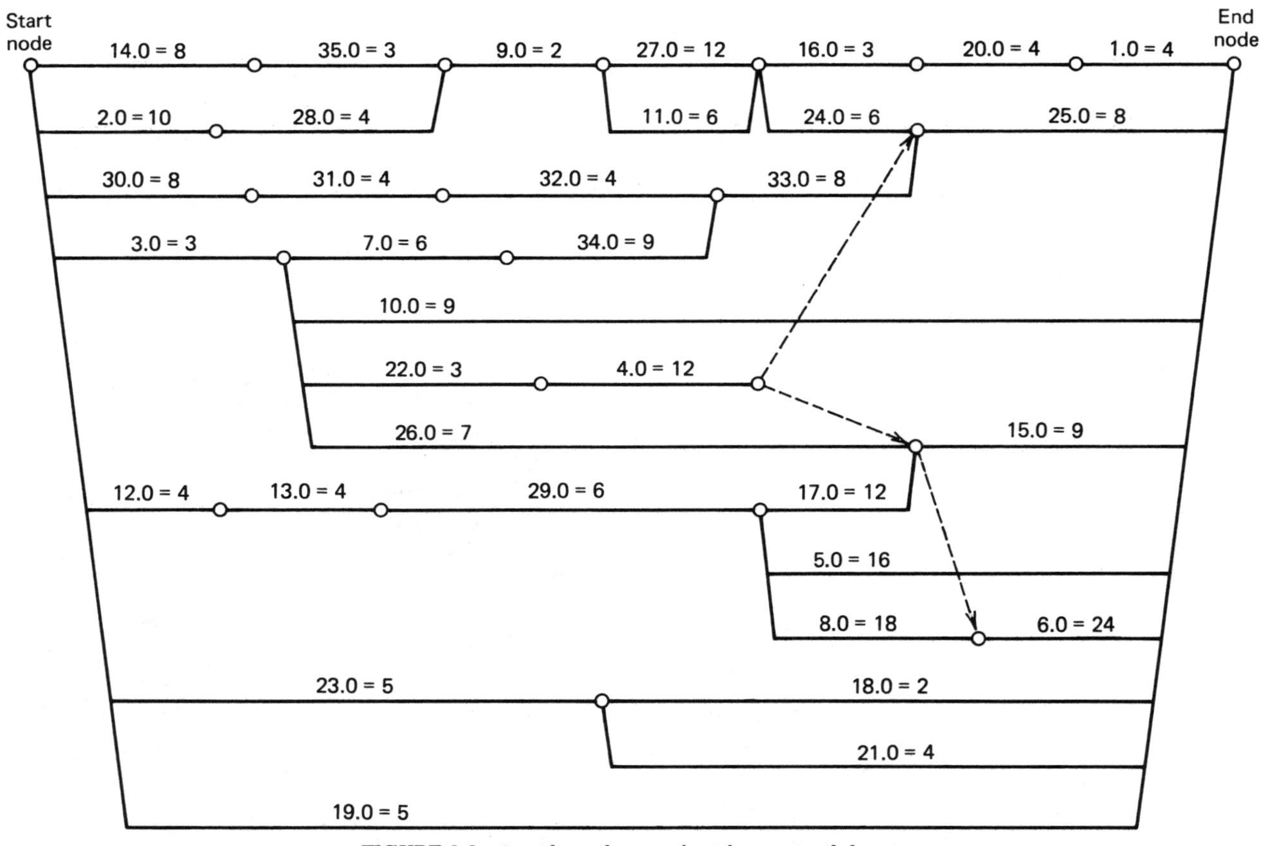

FIGURE 9.3 Sample goal network with associated durations

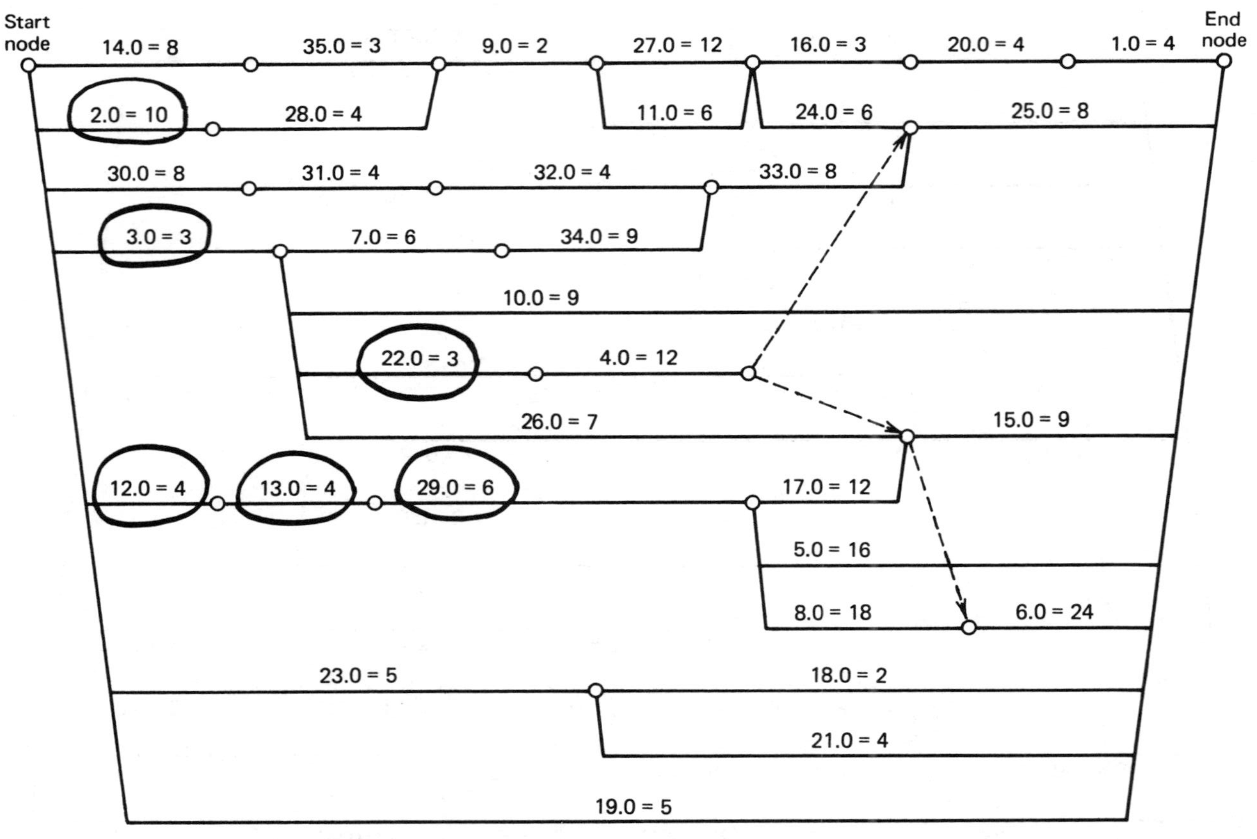

FIGURE 9.4 Sample goal network with goals selected for subdividing into tasks.

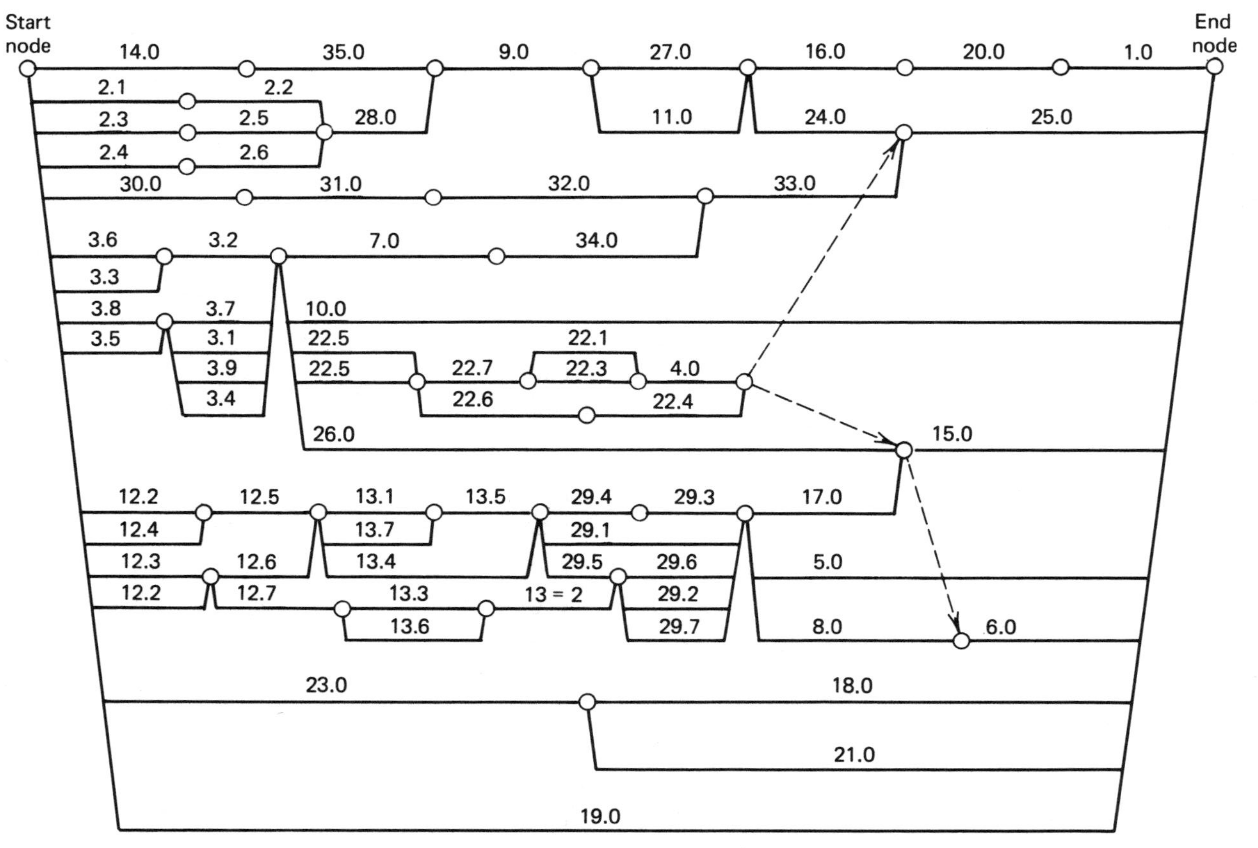

FIGURE 9.5 Sample task/goal network.

Goal Task Number	Planned Dates		Assigned Person	Department	Days Duration	Percentage of Time	Man-days
	Start	Finish					
1.0							
2.1							
2.2							
2.3							
2.4							
2.5							
2.6							
3.1							
3.2							
3.3							
3.4							
3.5							
3.6							
3.7							
3.8							
3.9							
4.0							
5.0							
6.0							
7.0							
8.0							
9.0							

FIGURE 9.6 The chart layout for duration estimating.

Goal Task Number	Planned Dates		Assigned Person	Department	Days Duration	Percentage of Time	Man-days
	Start	Finish					
1.0			C. Wood	Maintenance			
2.1			K. Brown	DP			
2.2			B. Beesley	Engineering			
2.3			J. Smith	Engineering			
2.4			S. Hinkleman	Materials			
2.5			J. Smith	Engineering			
2.6			K. Brown	DP			
3.1			S. Hinkleman	Materials			
3.2			K. Brown	DP			
3.3			B. Jones	Inventory Control			
3.4			R. Johnson	Production Control			
3.5			R. Johnson	Production Control			
3.6			R. Johnson	Production Control			
3.7			S. Weaver	DP			
3.8			B. Jones	Inventory Control			
3.9			S. Weaver	DP			
4.0			C. Wood	Maintenance			
5.0			B. Jones	Inventory Control			
6.0			B. Beesley	Engineering			
7.0			R. Johnson	Production Control			
8.0			J. Light	Marketing			
9.0			B. Black	VP Manufacturing			

FIGURE 9.7 First draft of personnel assignments.

Goal Task Number	Planned Dates Start	Finish	Assigned Person	Department	Days Duration	Percentage of Time	Mandays
1.0			C. Wood	Maintenance			
2.1			K. Brown	DP	3.0	0.5	1.5
2.2			B. Beesley	Engineering	6.0	1.0	6.0
2.3			J. Smith	Engineering	10.0	0.8	8.0
2.4			S. Hinkleman	Materials	14.0	1.0	14.0
2.5			J. Smith	Engineering	21.0	0.8	16.8
2.6			K. Brown	DP	9.0	0.5	4.5
3.1			S. Hinkleman	Materials	14.0	1.0	14.0
3.2			K. Brown	DP	2.0	0.5	1.0
3.3			B. Jones	Inventory Control	11.0	0.6	6.6
3.4			R. Johnson	Production Control	1.0	1.0	1.0
3.5			R. Johnson	Production Control	12.0	1.0	12.0
3.6			R. Johnson	Production Control	10.0	1.0	10.0
3.7			S. Weaver	DP	2.0	0.5	1.0
3.8			B. Jones	Inventory Control	4.0	0.6	2.4
3.9			S. Weaver	DP	2.0	0.5	1.0
4.0			C. Wood	Maintenance			
5.0			B. Jones	Inventory Control			
6.0			B. Beesley	Engineering			
7.0			R. Johnson	Production Control			
8.0			J. Light	Marketing			
9.0			B. Black	VP Manufacturing			

FIGURE 9.8 Estimated task durations and mandays.

Goal Task Number	Planned Dates Start	Finish	Assigned Person	Department	Days Duration	Percentage of Time	Man-days
1.0			C. Wood	Maintenance	88.0	0.5	44.0
2.1			K. Brown	DP	3.0	0.5	1.5
2.2			B. Beesley	Engineering	6.0	1.0	6.0
2.3			J. Smith	Engineering	10.0	0.8	8.0
2.4			S. Hinkleman	Materials	14.0	1.0	14.0
2.5			J. Smith	Engineering	21.0	0.8	16.8
2.6			K. Brown	DP	9.0	0.5	4.5
3.1			S. Hinkleman	Materials	14.0	1.0	14.0
3.2			K. Brown	DP	2.0	0.5	1.0
3.3			B. Jones	Inventory Control	11.0	0.6	6.6
3.4			R. Johnson	Production Control	1.0	1.0	1.0
3.5			R. Johnson	Production Control	12.0	1.0	12.0
3.6			R. Johnson	Production Control	10.0	1.0	10.0
3.7			S. Weaver	DP	2.0	0.5	1.0
3.8			B. Jones	Inventory Control	4.0	0.6	2.4
3.9			S. Weaver	DP	2.0	0.5	1.0
4.0			C. Wood	Maintenance	264.0	0.5	132.0
5.0			B. Jones	Inventory Control	352.0	0.5	176.0
6.0			B. Beesley	Engineering	528.0	0.5	264.0
7.0			R. Johnson	Production Control	132.0	0.5	66.0
8.0			J. Light	Marketing	396.0	0.5	198.0
9.0			B. Black	VP Manufacturing	44.0	0.5	22.0

FIGURE 9.9 Estimated task/goal durations and mandays.

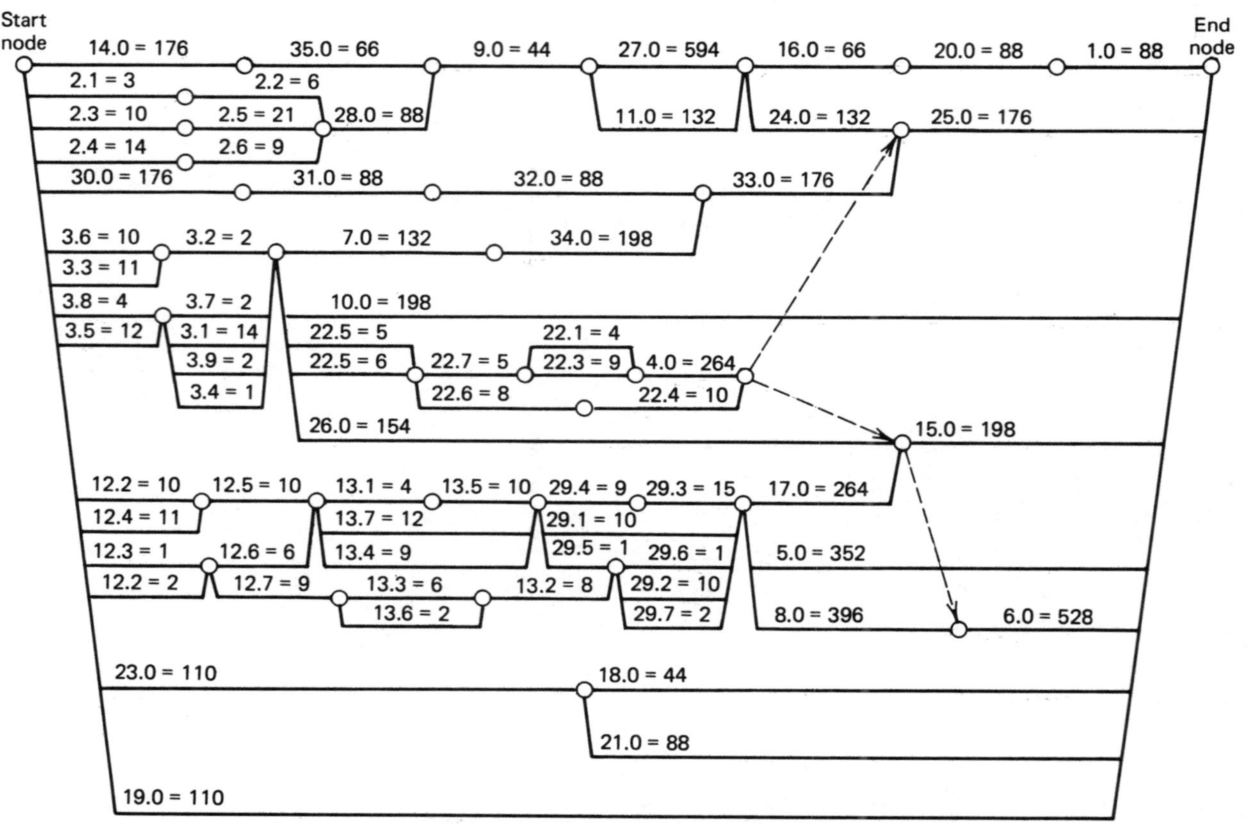

FIGURE 9.10 Sample task/goal network.

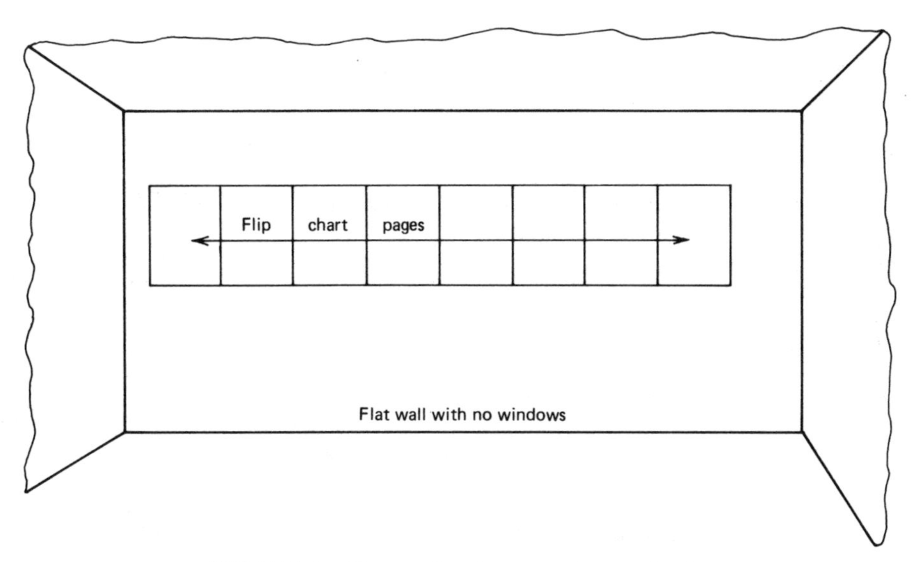

FIGURE 9.11. Hanging flip chart pages for road leveling.

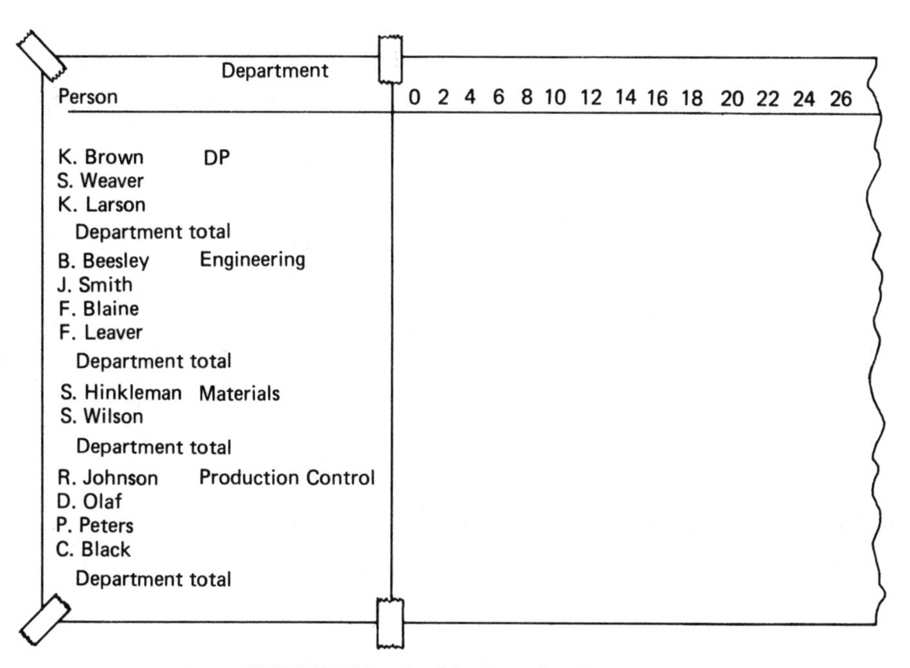

FIGURE 9.12. Load-leveling chart layout.

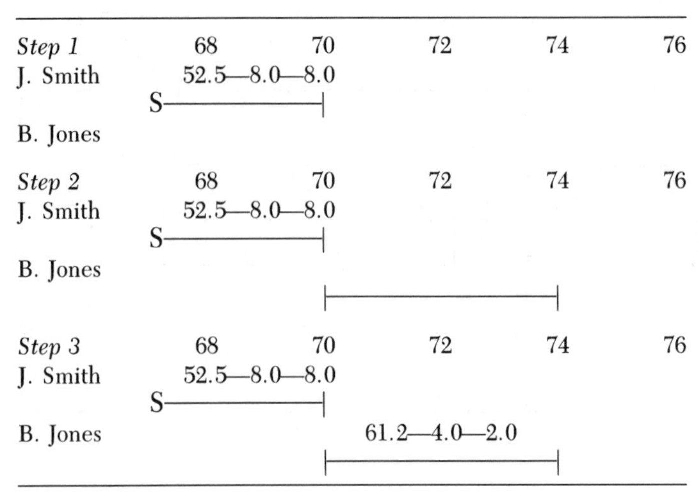

FIGURE 9.13 Plotting the tasks and goals.

		1983			1984		
Person	Department	2	3	4	1	2	3
K. Brown	DP	22.0	34.0	64.0	62.0	40.5	35.5
S. Weaver		10.5	18.5	22.0	30.0	28.5	30.0
K. Larson		2.0	6.5	7.0	7.0	6.5	6.0
Department total		34.5	59.0	93.0	99.0	75.5	71.5
B. Beesley	Engineering	40.0	42.0	48.0	54.5	58.5	60.0
J. Smith		38.0	43.5	47.5	50.0	59.0	59.5
F. Blaine		40.5	40.0	45.0	48.0	50.0	52.0
F. Leaver		45.0	50.0	48.0	66.0	50.5	48.0
Department total		163.5	175.5	188.5	21.5	218.0	219.5
S. Hinkleman	Materials	1.5	6.5	7.0	2.0	7.0	0
S. Wilson		3.0	3.0	1.5	0	0	1.5
Department total		4.5	9.5	8.5	2.0	7.0	1.5
R. Johnson	Production Control	4.0	4.0	42.0	60.0	61.0	59.5
D. Olaf		0	0	22.5	31.5	45.0	15.0
P. Peters		0	0	0	15.0	31.0	0
B. Black		42.0	0	0	0	0	0
Department total		46.0	4.0	64.5	106.5	137.0	74.5

FIGURE 9.14 Sample manday load chart by quarter.

K. Brown
19.2, 22.2, 22.3, 19.3, 1.4, 19.5, 22.4, 22.5, 22.6, 22.8, 22.9, 22.10, 22.11

S. Weaver
99.15, 99.46, 99.70, 99.35, 99.17, 99.29

K. Larson
99.13, 99.85, 99.134, 99.22, 99.62, 99.36, 99.135, 99.86, 99.136, 99.137, 99.87, 99.67, 99.68, 99.40, 99.68, 99.38, 99.147, 99.154, 99.157, 99.159, 99.131, 99.158, 99.160, 99.132

B. Beesley
99.163, 99.117, 99.48, 99.2, 4.1, 4.2, 99.123, 99.127, 99.141, 99.148, 99.18, 99.124, 99.146, 99.19, 99.56, 99.144, 99.151, 99.155, 99.145, 99.152, 99.153, 99.150

J. Smith
99.112, 00.113, 99.114

F. Blaine
99.120, 99.1, 99.11, 99.14

F. Leaver
99.57

S. Hinkleman
21.2, 21.2, 21.3, 21.6, 21.7, 21.4, 21.5, 21.8, 21.9, 21.10

FIGURE 9.15 Sample assigned person task sequence list.

Goal Task Number	Planned Dates		Assigned Person	Department	Days Duration	Percentage of time	Man days
	Start	Finish					
1.0			C. Wood	Maintenance	88.0	0.5	44.0
2.1			K. Brown	DP	3.0	0.5	1.5
2.2			B. Beesley	Engineering	6.0	1.0	6.0
2.3			J. Smith	Engineering	10.0	0.8	8.0
2.4			S. Hinkleman	Materials	14.0	1.0	14.0
2.5			J. Smith	Engineering	21.0	0.8	16.8
2.6			K. Brown	DP	9.0	0.5	4.5
3.1			S. Hinkleman	Materials	14.0	1.0	14.0
3.2			K. Brown	DP	2.0	0.5	1.0
3.3			B. Jones	Inventory Control	11.0	0.6	6.6
3.4			R. Johnson	Production Control	1.0	1.0	1.0
3.5			R. Johnson	Production Control	12.0	1.0	12.0
3.6			R. Johnson	Production Control	10.0	1.0	10.0
3.7			S. Weaver	DP	2.0	0.5	1.0
3.8			B. Jones	Inventory Control	4.0	0.6	2.4
3.9			S. Weaver	DP	2.0	0.5	1.0
4.0			C. Wood	Maintenance	264.0	0.5	132.0
5.0			B. Jones	Inventory Control	352.0	0.5	176.0
6.0			B. Beesley	Engineering	528.0	0.5	264.0
7.0			R. Johnson	Production Control	132.0	0.5	66.0
8.0			J. Light	Marketing	396.0	0.5	198.0
9.0			B. Black	VP Manufacturing	44.0	0.5	22.0

FIGURE 9.16 Sample assigned person/duration chart.

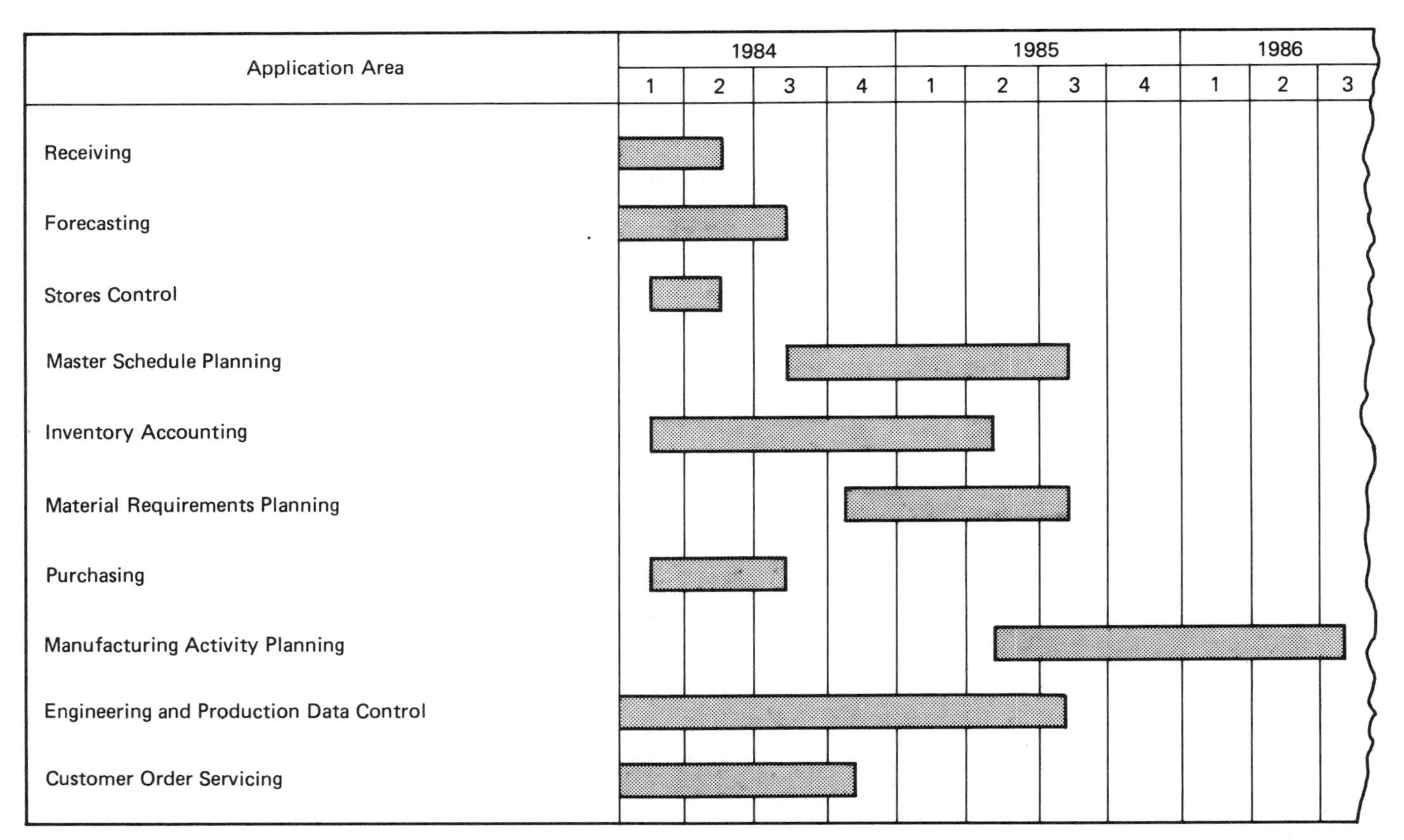

FIGURE 9.17 Sample project implementation summary chart.

For: D. Swartz Last Plan Update: 10/01/83

Dept: Engineering

Task/Goal Number	Planned Dates		Task/Goal Description	Planned Time		
	Start	Finish		Days Duration	Percentage of Time	Mandays
53.1	11/19/83	12/05/83	Design family routings based upon edits and potential family routing constructions for standard products, (assisted by B. Jones).	20.0	2.0	40.0
54.6	12/17/83	01/20/84	Review facilities and tooling numbering systems with projected requirements and usage.	20.0	1.0	20.0
52.4	12/19/83	01/22/84	Design family BOM's based upon edits and the potential family BOM constructions of standard products.	20.0	1.0	20.0
42.1	01/19/84	02/17/84	Test the product cost module against available data.	5.0	0.5	2.5
55.9	01/20/84	01/31/84	Load the facilities module with shop and tool data.	20.0	1.0	20.0
72.4	01/27/84	01/30/84	Set up engineering family bills of material for quoting.	3.0	1.0	3.0
71.2	01/27/84	01/31/84	Set up engineering data tables for use in routings for quoting.	4.0	1.0	4.0
56.3	02/17/84	03/10/84	Define the critical path through department facilities for quotation purposes having set up engineering data tables for use in the the facilities module for quoting.	15.0	1.0	15.0

FIGURE 9.18 Sample detail implementation page.

CHAPTER TEN

PHASE IV: JUSTIFYING THE SYSTEM

WHY JUSTIFICATION IS ESSENTIAL AND WAYS IN WHICH LACK OF IT CAN LEAD TO SYSTEM FAILURE

You have completed the study to this point and now your executive sponsor says, "It undoubtedly is intuitively obvious to the most casual observer that the implementation of this plan is justified and, therefore, we do not need to expend the time to formally justify it."

That's a compliment to you and the good job you've done. Suppose that your conclusions are:

I've done a good job.
The Company is satisfied.
I'm satisfied that this is a success.
I can cut this study short and move on to the next one.

Your conclusions may be right. On the other hand, these things could happen:

1. Your executive sponsor gets promoted to a corporate position. Her replacement is a firm believer that marketing drives the Company. He sets up new commission plans. He allocates money for sales bonuses. He invests in advertising. Guess where the money for these efforts is coming from? Your initial implementation costs, of course. He gets the money for his pet pro-

jects because it will increase sales by $X and only cost $Y. More important than your implementation plan? Nobody knows. You never figured out $X or $Y.

2. It is now three months after you put your plan together. Everything is on schedule. An unrelated critical emergency arises. By spending $Y, they can fix this emergency. The emergency is the requirement for a new warehouse, because the Company is out of space in the existing one. Of course, the $Y is your planned implementation money. Or, was it planned, since you did not do a justification? No, it wasn't planned, it was assumed. By the way, your implementation, in another three months, would have reduced inventory and eliminated the need for a new warehouse. Just try to prove that when you did not do a justification, did not commit the funds, and now someone screams crisis.
3. This is the same start as the above scenario. Your implementation plan is right on schedule but there is this guy named Billy Rubenstein out in the shop. He puts together a suggestion for management. He says that if they buy this multispindle mill for $Y, they can sell these other machines for $B, decrease maintenance costs by $C, and increase throughput in a bottleneck work center by 30%. This is a super idea. It will mean a savings of $50,000 per year, starting right now. What does your implementation plan save? Who knows? Where is the $Y money going to come from for the new mill? Of course, from the money that you assumed would be used for your implementation.

You will notice that in the nomination form in the appendix, I identify two types of studies that can be done. One is a full study, which includes all four phases. The other is an implementation planning study, which does not include Phase I, problem definition, or Phase IV, cost/benefit justification. You should not do an implementation planning study unless you are absolutely sure that the Company is totally committed to your proposed plan.

In summary, you do not have to do a study justification, but if you don't, you have substantially increased your potential for a study failure.

REVIEWING POTENTIAL BENEFIT AREAS TO EVALUATE

You completed problem definition, general system design, and implementation planning. Now you are going to justify your plan. You'll lose a few hours of effective study time, because this is a new area and requires a shift in thinking.

Tell your team that you will discuss two key areas, benefits and costs. You will deal with benefits first, so tell your team not to get costs involved in the discussion until you get to that area.

A benefit is a general term. It may be tangible, meaning that you can quantify it in monetary terms. It may be intangible, meaning that it is impossible to assign a dollar value to the worth of the benefit.

When you start this phase, do not get hung up on whether a benefit is tangible or intangible. Begin by going through your functional relationship chart and your general application design. For each function/area, make a list on a flip chart of potential impact areas.

Your team will be slow in starting this process due to the necessary "shift in thinking." Use a foil such as the one in Figure 10.1 to get their thinking started. They will not use all of the potential impact areas on your foil.

The examples in Figure 10.1 are not intended to be all inclusive. They are meant to stimulate your team into thinking of the potential benefit areas that exist in their company. As you step your team through the areas in Figure 10.1, add the potential benefits to your flip chart list.

HOW TO DOCUMENT AND SUMMARIZE TANGIBLE BENEFITS

You just completed a flip chart that lists potential benefit areas that "might" be considered in a cost/benefit analysis. You will now want to evaluate each of the potential areas and determine if they can be quantified in dollars. If an item can be quantified, then it is a tangible benefit. If it can't, then it is an intangible benefit.

First, frame a blank flipchart with the headings shown in Figure 10.2. Assume that someone on your team suggested that overall productivity might be increased when the planned systems are implemented. You ask, "For how many people?" They answer, "973." You ask, "What is an average annual salary?" A team member leaves the room to find out. He comes back and says, "$16,500." Now you need the potential percentages for productivity improvement. You ask, "What is the most likely or conservative estimate of the productivity improvement percentage that will occur?" Your team says, "10%." That is the low end of the scale. Now look for the high end. You ask, "If everything in the world went right, what is the maximum possible improvement percentage?" They answer, "30%." Now all that you lack is the medium range or the probable percentage. Usually this will be an average of the high and low ranges, or in this example, 20%. Note that this is not always true. Sometimes your team will want to weight the percentages toward the low or high end. Also, sometimes you may not get a percentage range, but have a fixed quantity.

Now on your blank (but framed) flip chart, document what you just learned about productivity improvement. Your flip chart should look like Figure 10.3. Leave the far right column (the chosen quarterly amount) blank at this time.

Complete all of your tangible benefits in the same manner as for the productivity improvement example. Your completed tangible benefits could look like Figure 10.4.

At this point in your study, you should review your tangible benefit detail charts with your Executive Sponsor and his or her financial people. Do they agree with the benefit descriptions? Are there any benefits that the study team missed? Are the percentages realistic? And finally, of the three potential attainments (likely, probable, and possible), which one, for each specific benefit, does management feel most comfortable with? As they select a column for a benefit, divide the annual amount by four and enter that quantity in the chosen Quarterly Amount column. Note that one column (as probable) does not necessarily apply to all benefits. For each individual benefit, management should select the column (amount) that they feel is reasonable.

When the above review is completed, your Tangible Benefit Detail charts should look like Figure 10.5.

You now want to summarize, or total, your benefits. You cannot simply total the amounts that you have identified for two reasons:

1. Different benefits start at different times.
2. The amount of benefit applicable by quarter will vary until the maximum amount is achieved.

Construct a flip chart page with the following headings:

Benefit Area	Total Quarterly Benefit	Date Benefit Starts	Date Benefit Totally Effective	1984				1985			
				1	2	3	4	1	2	3	4

Enter a summarized description of the benefit into the benefit area. Enter the total quarterly benefit from your tangible benefit detail charts into the total quarterly benefit area.

You still have your load-leveling chart (from the implementation planning phase) on your wall. It contains all the tasks and goals that you defined, plus the start–stop dates. Look at each benefit and, based on the tasks that relate to the benefit, decide in which quarter of which year the benefit will start and some of the benefit dollars will be realized. Also identify when the benefit will be totally effective. Enter these dates on your flip chart.

An S curve is a useful tool for deciding how much of the benefit amount is applicable for a quarter. Review Figure 10.6. The height of the S curve is dependent on the maximum amount of the benefit. The slope of the curve is dependent on the number of quarters between when the benefit starts and when it is totally effective.

The S curve is only a guide. It provides you with a way of simulating the gradual realization of benefits. It allows you to show a slow buildup, then a steady increase, and finally a small accumulation of the fringe areas that makes the benefit totally effective. Once you have some base numbers representing the benefit dollars by quarter, adjust them to what your study team feels is the most realistic view.

You should now have a chart similar to Figure 10.7. Note that Figure 10.7 shows only a one-year benefit plan. Your chart should cover the entire horizon of your plan—five to ten years.

HOW TO DEFINE AND VALIDATE INTANGIBLE BENEFITS, AND WHY THEY ARE IMPORTANT

When you started on benefits, you made up a flip chart of potential benefits. If you have been crossing off these potential benefits that you determined to be tangible benefits, you will probably find that you still have 10 to 20 items left on the flip chart after you are done. These are the benefit areas that no one feels can be quantified. These are your intangible benefits.

Of the 10 to 20 remaining intangible benefits, cross off those that:

Top management would question as really being benefits

Are insignificant or minor

Try to summarize or restate your remaining intangible benefits so you have no less than three and no more than six. Resequence the group so that the ones with the most interest to top management appear at the top of the list. You should now have a flip chart similar to Figure 10.8.

You may be tempted to spend very little time to summarize intangible benefits. Believe me, it is worth the time. Allow me to give you an example. My study team was doing the final presentation. There were about 20 managers in the room. The company president (who was not the Executive Sponsor) sat quietly while my team went through the tangible benefits. Then they put up one foil showing five intangible benefits. The president's comment was, "If you think that you have a shot at that first item (it was improving customer service), then I have no concern about the numbers on your tangible benefits. This one item will justify the system by driving the tangible benefits."

If that situation happens to you only once out of five studies, it is still worth the time to define and illustrate intangible benefits.

You have now completed your estimation of benefits. Next, we identify costs and then relate benefits and costs to each other.

REVIEWING POTENTIAL COST AREAS TO EVALUATE

When you started to address benefits, your study team had to go through a shift in thinking. You will encounter the same reaction when you want your team to stop thinking about benefits and start thinking about costs.

Once that you identify the cost areas that you want to evaluate, you should make specific assignments to each study team member for the preparation of the cost data. Now, of course, different people think in different levels of detail. You, however, would like your final cost picture to reflect a similar level of detail throughout all of the cost areas. Therefore, during the start-up phase, you will have to set the level of detail that is desired. Let's consider some examples.

Your company has decided to buy the required computer software for their manufacturing systems instead of designing their own. The cost is $230,000. The computer that your company is acquiring will be installed in the fourth quarter of 1983. Therefore, the software cost is also placed in the fourth quarter of 1983.

This is too general an approach. The $230,000 or more worth of software actually consists of 20 program packages, some relating to computer system operation and some relating to the manufacturing applications. Most of the computer system operation software will be required when the computer is installed. The manufacturing application software will be phased in. Phased in when? Your load-leveling chart is still on the wall. You know from your tasks (on the load-leveling chart) when the bill of material processing software is required. It is not at the same time as when the material requirements planning software is required, since bills of material must be structured and validated before material requirements planning can be effective. Therefore, depict the software cost at the point when it will occur according to your load-leveling chart.

In another example, a team member is assigned the responsibility to define education costs. She completes the chart shown in Figure 10.9. (In actuality, she completes eight charts similar to the one in Figure 10.9.) The information is very well organized. The team member obviously put a lot of thought into the preparation of these education costs, but let's put it in perspective. In the benefit area, we may plan to reduce a $6.1 million manufacturing inventory by about 30%. In contrast, B. Jones is scheduled to attend a one-day Communication Concepts and Facilities course at the plant in March 1984 at a cost of $7. Who cares?

What you really need is an estimate of how many people will require what types of education in what quarter of what year. Estimate some tuition, travel, and living expenses that will apply on the average and multiply the dollar amount by the number of potential students. Now increase your education costs by 50% because education always costs more than what you originally thought—even if you had the detail of Figure 10.9.

Set the level of detail that you desire from your study team with examples similar to those that I used.

Now you have to determine what cost areas you are going to evaluate. Figure 10.10 can be used to start the thinking process of your team. Follow the same initial steps that you used when you started to define benefits. Note that Figure 10.10 illustrates some potential cost areas that you may not consider to be part of a manufacturing information control system implementation, but you should recognize that the cost of any and all effects on the company as a result of your study should be estimated. If you do not do it, then the company's Vice President of Finance will insist on it, so you may as well do it initially.

When you reach this point, you should have a flip chart with a list of Cost Areas that are applicable to this company's study. Assign (ask for volunteers) the cost areas to your team members. Two-person teams work well for this phase. Their purpose is to document the cost requirements.

HOW TO DOCUMENT COST REQUIREMENTS

The documentation of costs is a two-step process. First the requirements need to be identified, and then the costs need to be spread by yearly quarter. Figure 10.11 illustrates the way that one company identified their software requirements and spread those costs.

To prepare the chart in Figure 10.11, first the software packages and their costs were identified. The second step was to review the load-level chart and determine the approximate install date of the package. Notice that during the first quarter of the installation of a package, the lease cost is not necessarily equal to three times the monthly rate. You may plan for the package to be installed during the last month of the quarter and therefore only one month of lease cost appears in the quarter.

Your study team should develop charts similar to Figure 10.11 for every cost area. These should be reviewed by the entire team and total agreement obtained that all applicable costs have been identified.

Finally, to wrap up the cost area, you need to summarize all of your costs on a single page as you did with benefits when you prepared the benefits by quarter summary. This chart, as you might expect, is called the costs by quarter summary. Figure 10.12 is an example of the format to be used. Note that although Figure 10.12 illustrates a two-year plan, your chart should reflect the full duration of your plan, either five or eight years.

The data that you have derived for benefits and costs are used in the next section to perform a break-even analysis.

PREPARING A COST-BENEFIT SUMMARY

The two charts that you prepare in this section will wrap up the phase on justifying the system.

First, review Figure 10.13. Now frame a similar flip chart with the number of yearly quarters in your plan. Enter the data under Gross Benefits from your benefits by quarter summary chart. Enter the data under costs from your costs by quarter summary chart. Notice that costs are entered as negative amounts because they represent a cash outflow.

Second, subtract the costs from the benefits and you obtain the net benefits. Note, however, that just because the net benefits go positive in the third quarter does not mean that the implementation has recovered expended costs. You need to calculate a cumulative total of the net benefits. Now you can see that by the second quarter of the following year, all previous costs have been recovered and now benefits exceed continuing costs.

Although many people easily understand the set of numbers that you have just developed, it normally can be illustrated more effectively if you construct the line chart shown in Figure 10.14. This chart clearly shows the relationships of the line items of data that you have calculated.

You have now completed the phase on Justifying the System. This is an ideal time for an interim review. Pass the numbers by your company's management. Obtain their agreement now so that there are no surprises during the final presentation.

A note of caution—it is helpful to know, prior to starting this phase, what an acceptable break-even point would be for your specific company. Acceptable break-even points vary by company and also vary based on economic trends.

You have almost completed your study. All that remains is to define how the implementation will be managed, the preparation of the Final Report, and the Final Presentation.

FIGURES

Key Functions	Potential Impact Areas
1. Business Planning	Improved profit analysis Investment determination Product emphasis/analysis
2. Forecasting	Improved sales analysis by region Building forecast models by sales family
3. Demand Analysis	Improved techniques for blending actual sales to forecasts
4. Production Planning	Ability to aggregate individual items to product families for easier analysis Resource testing of product families (hours/costs/sales)
5. Master Schedule Planning	More accurately provide promised ship dates Reduced reschedules caused by out of capacity conditions Reduction in finished goods inventory
6. Order Entry and Inquiry	Increased ability to meet promised ship dates Reduced rework by decreasing initial configuration errors Reduction in back orders
7. Bills of Material	Decrease engineering cost for error correction Decrease the time to process engineering changes Reduction in obsolete/surplus inventory Improved ability to cost products
8. Inventory Accounting	Improved inventory accuracy Improved security Decreased shortages Decreased inventory by physical staging elimination
9. Material Requirements Planning	Decreased purchased component/raw material inventory Improved planning accuracy Ability to establish long term supplier agreements
10. Capacity Requirements Planning	Reduced number of missed shop order due dates Decreased work in process inventory
11. Shop Order Release	Ability to work on the right order at the right time Decreased work in process inventory
12. Shop Floor Feedback	Reduced reaction time on rework Ability to de-expedite an order Improved order status knowledge Rapid identification of scrap origination Decreased production count errors
13. Tools and Facilities	Improved tool availability Ability to maintain (know) tool usage and relate it to repair or replacement Knowledge of tool maintenance costs
14. Routings	Improved standards on hours/costs Ability to analyze labor effectivity

FIGURE 10.1 Potential benefit areas.

Key Functions	Potential Impact Areas
15. Purchasing	Improved purchase order control Ability for effective vendor analysis Reduction in expedited orders
16. Receiving	Decreased cycle time from receiving to stores Increased ability to receive what was ordered
17. Plant Maintenance	Decreased remedial maintenance Improved preventive maintenance scheduling Increased machine life Decreased line shutdowns
18. Product Costing	Ability to rapidly prepare a standard/actual cost variance Reduced cycle time to react to cost changes Increased ease of costing products when a component "replace where used" has been performed.

FIGURE 10.1. *(continued)*

Benefit Description	Potential Annual Attainment (in thousands of dollars)			Chosen Quarterly Amount
	Likely	Probable	Possible	

FIGURE 10.2 Framing the tangible benefit analysis.

Benefit Description	Potential Annual Attainment (in thousands of dollars)			Chosen Quarterly Amount
	Likely	Probable	Possible	
1. Through the installation of manufacturing systems, overall productivity will be increased.				
973 employees, × an average salary of $16,500,				
× a 10% increase =	$1605			
× a 20% increase =		$3211		
× a 30% increase =			$4816	

FIGURE 10.3 Starting the tangible benefit analysis.

Benefit Description	Potential Annual Attainment (in thousands of dollars)			Chosen Quarterly Amount (in thousands of dollars)
	Likely	Probable	Possible	
1. Through the installation of manufacturing systems, overall productivity will be increased.				
973 employees, × an average salary of $16,500,				
× a 10% increase =	1605			
× a 20% increase =		3211		
× a 30% increase =			4816	
2. Scrap and rework will be reduced due to better receiving inspection procedures, having the proper equipment installed, improved quality control and a training program.				
$800,000 scrap and rework annually				
× a 25% reduction =	200			
× a 30% reduction =		240		
× a 35% reduction =			280	
3. Inventories will be reduced due to building what we need, reducing obsolete parts, utilizing stocked parts, utilizing stocked parts with a good locator system, improved inventory accuracy, use of an improved vendor analysis, material requirements planning, installation of capacity planning and tool inventory control.				
Inventories				
$4,708,000—WIP				
$1,500,000—Tool (est.)				
$6,150,000—In stock (Manufacturing)				
$5,550,000—In stock (Purchasing)				

FIGURE 10.4 A sample of the tangible benefit analysis.

Benefit Description	Potential Annual Attainment (in thousands of dollars)			Chosen Quarterly Amount (in thousands of dollars)
	Likely	Probable	Possible	
Carry cost = 20%				
20% × $4,708,000 WIP				
× 10% =	$ 94			
× 15% =		$ 141		
× 20% =			$ 188	
20% × $1,500,000 Tool				
× 3% =	9			
× 5% =		15		
× 8% =			24	
20% × $6,150,000 Manufacturing				
× 20% =	246			
× 30% =		369		
× 50% =			615	
20% × $5,550,000 Purchasing				
× 10% =	$ 111			
× 17% =		$ 188		
× 25% =			$ 278	
4. By decreasing lead times, we will improve our competitive position and increase our gross sales margin				
25% margin × $39,000,000 gross sales margin				
× 5% =	$ 488			
× 7% =		$ 683		
× 10% =			$ 975	
5. By shipping on time, we will increase our gross sales margin of business currently lost on specific products				
17% margin × $30,000,000 gross sales margin				
× 5% =	255			
× 7% =		357		
× 10% =			510	
6. On time shipments of service parts will increase the service parts business				

Benefit Description	Potential Annual Attainment (in thousands of dollars) Likely	Probable	Possible	Chosen Quarterly Amount (in thousands of dollars)
50% margin × $9,000,000 service parts				
× 5% =	225			
× 10% =		450		
× 15% =			675	
7. Through better receiving procedures, an accurate locator system and inventory transactions, the need for an annual physical inventory will be eliminated. Cost of an annual physical is $25,000.	25	25	25	
8. On-line systems will reduce paperwork and office work through time, resulting in reduced clerical personnel and cost avoidance in hiring additional personnel for the future.				
Base salary of $13,000				
× 11 people =	143			
× 16 people =		208		
× 20 people =			260	
9. Through the new joint material requirements planning and capacity systems, backorders will be reduced. Cost to process a backorder = $17				
2,000 backorders × $17 = $34,000				
× 25% =	9			
× 30% =		10		
× 50% =			17	

FIGURE 10.4 *(continued)*

Benefit Description	Potential Annual Attainment (in thousands of dollars) Likely	Probable	Possible	Chosen Quarterly Amount (in thousands of dollars)
1. Through the installation of manufacturing systems, overall productivity will be increased.				
973 employees, × an average salary of $16,500,				
× a 10% increase =	1605			
× a 20% increase =		3211		803
× a 30% increase =			4816	
2. Scrap and rework will be reduced due to better receiving inspection procedures, having the proper equipment installed, improved quality control and a training program.				
$800,000 scrap and rework annually				
× a 25% reduction =	200			
× a 30% reduction =		240		60
× a 35% reduction =			280	
3. Inventories will be reduced due to building what we need, reducing obsolete parts, utilizing stocked parts, utilizing stocked parts with a good locator system, improved inventory accuracy, use of an improved vendor analysis, material requirements planning, and the installation of capacity and tool inventory control.				
Inventories				
$4,708,000—WIP				
$1,500,000—Tool (est.)				
$6,150,000—In stock (Manufacturing)				
$5,550,000—In stock (Purchasing)				

Benefit Description	Potential Annual Attainment (in thousands of dollars) Likely	Probable	Possible	Chosen Quarterly Amount (in thousands of dollars)
Carry cost = 20%				
20% × $4,708,000 WIP				
× 10% =	94			
× 15% =		141		35
× 20% =			188	
20% × $1,500,000 Tool				
× 3% =	9			2
× 5% =		15		
× 8% =			24	
20% × $6,150,000 Manufacturing				
× 20% =	246			
× 30% =		369		92
× 50% =			615	
20% × $5,550,000 Purchasing				
× 10% =	111			
× 17% =		188		
× 25% =			278	70
4. By decreasing lead times, we will improve our competitive position and increase our gross sales margin				
25% margin × $39,000,000 gross sales margin				
× 5% =	488			
× 7% =		683		171
× 10% =			975	
5. By shipping on time, we will increase our gross sales margin of business currently lost on specific products				
17% margin × $30,000,000 gross sales margin				
× 5% =	255			
× 7% =		357		89
× 10% =			510	
6. On-time shipments of service parts will increase the service parts business				

FIGURE 10.5 Completed sample tangible benefit analysis.

Benefit Description	Potential Annual Attainment (in thousands of dollars) Likely	Probable	Possible	Chosen Quarterly Amount (in thousands of dollars)
50% margin × $9,000,000 service parts				
× 5% =	225			
× 10% =		450		113
× 15% =			675	
7. Through better receiving procedures, an accurate locator system and inventory transactions, the need for an annual physical inventory will be eliminated. Cost of an annual physical is $25,000.	25	25	25	25
8. On-line systems will reduce paperwork and office work through time, resulting in reduced clerical personnel and cost avoidance in hiring additional personnel for the future.				
Base salary of $13,000				
× 11 people =	143			36
× 16 people =		208		
× 20 people =			260	
9. Through the new joint material requirements planning and capacity systems, back orders will be reduced. Cost to process a back order = $17				
2,000 back orders × $17 = $34,000				
× 25% =	9			
× 30% =		10		
× 50% =			17	4

FIGURE 10.5 *(continued)*

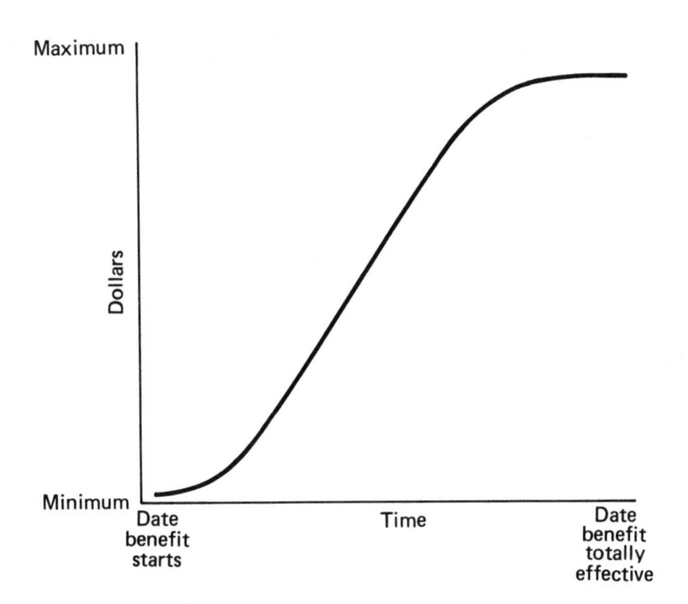

FIGURE 10.6 S Curve for benefit estimating.

Benefit Area	Total Quarterly Benefit	Date Benefit Starts	Date Benefit Totally Effective	Quarterly Benefits for Year 1984 (in thousands of dollars)			
				1	2	3	4
1. Personnel productivity	803	2/84	4/86		70	125	225
2. Reduced scrap and rework	60	1/84	4/84	7	25	50	60
3. Reduced inventories							
A. WIP inventory	35	3/84	4/85			5	9
B. Tool inventory	2	2/84	4/85		1	1	1
C. Manufacturing in-stock inventory	92	2/84	1/86		8	15	35
D. Purchasing in-stock inventory	70	1/84	4/85	7	12	20	28
4. Competitive position and gross sales increase profit	171	4/84	4/87				20
5. Recapture lost sales profit	89	4/84	3/87				12
6. Service part sales increase profit	113	4/84	2/86				15
7. Physical inventory	25	4/84	4/84				25
8. Clerical personnel reduced	36	2/85	2/87				
9. Backorders	4	4/84	2/87				1
Total benefits by quarter				14	116	216	431

FIGURE 10.7 Summary of benefits by quarter.

1. Better customer service will be provided through improved forecasts and better scheduling of production capacities to satisfy actual ship-date requirements.
2. Through the capacity planning network of order status information, customer inquiries can be answered in a timely, accurate manner.
3. Long-range capacity requirements planning will provide the capability to perform facility planning.
4. By improved definition, collection, and distribution of costs by product, management will be better able to make decisions regarding prices, product lines, equipment replacement, and make-or-buy decisions.

FIGURE 10.8 Sample intangible benefits.

Course/Education Required	Attendee	Date Required	Duration	Location	Costs			
					Tuition	Travel	Living	Total
Data Base Concepts and Facilities	B. Jones	03/84	1 day	here	—	—	—	—
Data Base Application Programming	J. Smith	03/84	3 days	Chicago	357	185	225	767
Data Base Application Analysis	J. Doe	03/84	5 days	Chicago	435	185	375	995
Data Base Implementation I	J. Doe	03/84	5 days	Chicago	635	185	375	1195
Data Base Implementation II	J. Doe	04/84	4 days	Chicago	393	185	300	878
Data Base/Communications Interface	B. James	04/84	14 hours	here	375	—	—	375
Data Dictionary System	B. James	04/84	3 days	Chicago	330	185	225	740
Data Base Design and Administration	J. Doe	03/84	10 days	N.Y.C.	1107	250	750	2107
Data Base Structures	J. Doe	04/84	2 days	Chicago	270	185	150	605
Bridge Implementation	J. Doe	01/85	2 days	Chicago	285	185	150	620
Communication Concepts/Facilities	B. Jones	03/84	1 day	here	7	—	—	7
Communication Command Level Coding	J. Smith	04/84	4 days	Chicago	525	185	300	1010
Communication Advanced Programming	J. Smith	04/84	4 days	L.A.	515	300	300	1115
Communication Text vs. High Level	J. Smith	03/84	1 day	Chicago	—	185	75	260
Communication Installation Workshop	B. James	04/84	5 days	Chicago	735	185	375	1295
Communication Performance Analysis	B. James	04/84	5 days	Chicago	855	185	375	1415

FIGURE 10.9 Excessive education cost detail.

Cost Areas	Types of Costs
1. Education	Data processing classes
	Manufacturing classes (generic)
	Professional society conferences
	Video tapes, manuals, books
	Executive classes
	Terminal usage training
	Application usage training
2. Data processing software	System software
	Application software
	Distributed system control software
	(on above, consider lease/purchase)
3. Data processing equipment	Central site host equipment
	Distributed site equipment
	Intelligent work stations
	Nonintelligent terminals
	Customized shop floor terminals
4. Manufacturing equipment	Analog/digital interface equipment
	Robots
	Load cells
	Numerical control machines
	Forklifts
	Conveyor systems
	Specialized maintenance equipment
5. Labor	Programmers
	Analysts
	Interface analysts (Mfg/DP)
	Terminal operators
	Implementation project manager
	Training coordinator
	Other (Manufacturing, Engineering, Management)
6. Miscellaneous costs	Video tape/disc player for education
	Set-up of a training room
	Office furniture
	Foremen offices on the shop floor
	Equipment freight costs
	Modem/cabling installation
	Leasing company buyout
	Communication line fee
	Personnel recruiting costs
	Video tape lending library costs
	Conversion costs (on DP equipment)

FIGURE 10.10 Cost areas for evaluation.

Description	Monthly Cost	Install Date (quarter/yr.)	1984				1985				1986		
			1	2	3	4	1	2	3	4	1	2	3
DOS ADV. FUNCTION	200	4/84				200	600	600	600	600	600	600	600
DOS PT	175	1/85					350	525	525	525	525	525	525
COBOL	140	4/84				280	420	420	420	420	420	420	420
SORT MERGE	150	4/84				300	450	450	450	450	450	450	450
DB DOS	395	4/84				790	1185	1185	1185	1185	1185	1185	1185
DC DOS	475	4/84				950	1425	1425	1425	1425	1425	1425	1425
MSSG. CTL	300	4/85								900	900	900	900
MSSG. CTL/DMS	125	4/85								125	375	375	375
DATA DICTIONARY	290	4/85								290	870	870	870
ETSS II	250	1/85					500	750	750	750	750	750	750
DMS/DOS	360	1/85					360	1080	1080	1080	1080	1080	1080
ATMS II	590	4/85								590	1770	1770	1770
BOM BATCH	50	1/85					150	150	150	150	150	150	150
BOM ON LINE	95	1/85					285	285	285	285	285	285	285
MRP	105	4/85								105	315	315	315
PROD. COST	115	4/85								115	345	345	345
INV. ACCTG	70	2/85						140	210	210	210	210	210
SHOP ORDER REL.	150	4/85								150	450	450	450
ROUTINGS	140	2/85						140	420	420	420	420	420
FACILITIES	140	1/85					140	420	420	420	420	420	420
FORECASTING	150	1/85					300	450	450	450	450	450	450
CRP	400	2/86										400	1200
Totals						2520	6165	8020	8370	10645	13395	13795	14595

FIGURE 10.11 Software costs by quarter.

	Dollars in Thousands							
	19XX				19XX			
Cost Area	1	2	3	4	1	2	3	4
1. Education	2.9	0.7	4.9	4.9	3.0	2.5	2.0	1.5
2. Data processing software	5.5	5.9	5.8	6.2	6.2	6.3	6.3	6.4
3. Data processing equipment	68.1	70.2	80.2	100.1	100.1	100.9	101.1	101.1
4. Manufacturing equipment	70.0	10.0	50.0	2.0	5.0	60.0	20.0	10.0
5. Labor	80.0	80.0	100.0	100.0	120.0	120.0	140.0	140.0
6. Miscellaneous costs	62.5	386.5	142.5	56.8	21.0	11.5	22.0	75.0
Total costs	289.0	553.3	383.4	270.0	255.3	301.2	291.4	334.0

FIGURE 10.12 Summary of costs by quarter.

	19XX	19XX				19XX			
	4	1	2	3	4	1	2	3	4
Gross benefits	0	12.0	125.0	155.8	188.3	231.3	330.6	375.3	441.3
Costs	−84.9	−122.1	−143.0	−142.0	−145.7	−150.3	−159.8	−152.1	−169.3
Net benefits	−84.9	−110.1	− 18.0	13.8	42.6	81.0	170.8	223.2	272.0
Cumulative net benefits	−84.9	−195.0	−213.0	−199.2	−156.6	− 75.6	95.2	318.4	590.4

FIGURE 10.13 Break-even calculation.

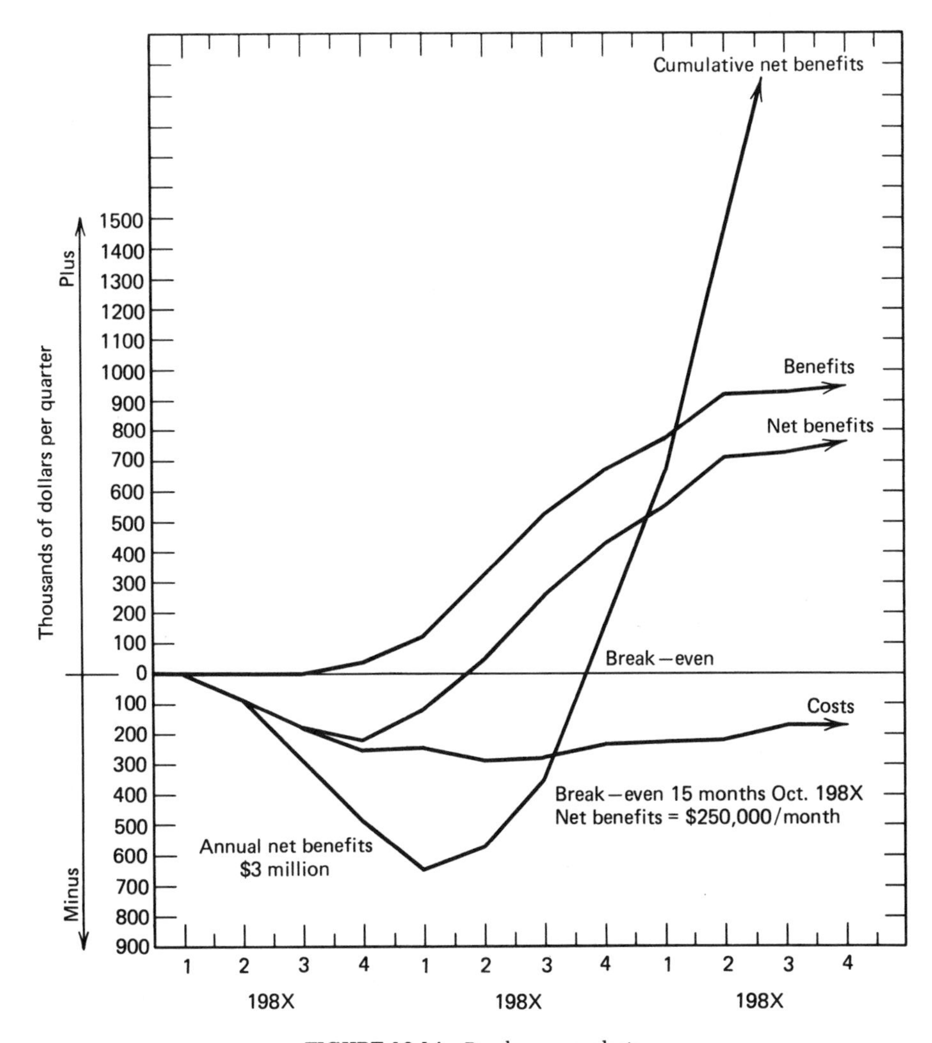

FIGURE 10.14 Break-even analysis.

PART FOUR

FINAL STEPS

The Birth of a System

CHAPTER ELEVEN

PREPARING FOR EFFECTIVE PROJECT MANAGEMENT

OBJECTIVES FOR PROJECT MANAGEMENT

There is really only one key objective relating to project management. Build the project tracking and controls into your plan necessary to ensure its implementation. All too often, we assume that the newly appointed project manager knows what he or she is supposed to do. This assumption is usually wrong.

Your project manager, without guidance, could decide to track activity on tasks by jotting notes on the implementation planning pages, such as "Late start," "50% complete," or "Completed on XX/XX/XX." While the notes are informative for each task, this approach is difficult to summarize for top management review.

The ability to summarize task performance on a single page and actual cost benefit data on another page is the key control to an effective project management system. The summarization should also include current projected completion status. When the president asks how a slippage will affect the completion date, your project manager should be able to answer that the completion date has slipped from, say, January 16, 19XX to February 14, 19XX.

To define how project management is to be conducted, two key elements are required: the project manager's job description and the tracking documents.

HOW TO DEVELOP A PROJECT MANAGER'S JOB DESCRIPTION

When I get to this section of a study, I use Figure 11.1 as a foil on the overhead projector to establish an example that can be expanded.

No one accepts the Figure 11.1 example as their final job description. It should be used as a starting point. Let's review some of the areas that promote discussion.

No one can disagree with the principal function or the general responsibilities. These are pure motherhood statements. They are included, however, to ensure that the other managers in the company understand the broad scope of the project management function.

Under specific responsibilities, Section 1.A is a must. This cannot be deleted. It is mandatory that the weekly reviews take place to ensure that task status is known and potential problems can be resolved. Some deviation might occur with Section 1.B, the top management monthly reviews. Some companies want to specify the top management individuals by name. Some want to change the frequency to twice a month. Note that a frequency of "upon demand" is not acceptable. Unless you have set up a specific schedule, the meetings will not take place.

Most people agree with Section 2.A and 2.B, since this is the major responsibility of the project manager. You may get some objections to Section 2.C. The most frequent comment is "I, as project manager, should not have to resolve nonperformance of tasks. I am not high enough in the organization to enforce performance." Your problem is that your project manager has responsibility with no authority. He does not have any clout. If you accept this situation, your plan will not be installed as scheduled.

Section 3 is completely omitted from the job description about 60% of the time, for a number of reasons including:

1. To track costs and benefits, a starting point is required. We think that we can reduce inventories by 20%, but we do not know the value of our inventories now. We cannot track against an unknown starting point.
2. We did the Justification Phase and showed a break-even period of 14 months. We are satisfied with that.
3. Too many extrinsic variables can affect cost benefit tracking, such as changing interest rates, changed carry cost rates, products added and deleted to the total product line, and so forth.
4. It is not necessary to track costs and benefits. If this project is not successfully installed, we'll be out of business.

As you can see, your customer may have one or more of a multitude of reasons for not including this section. I wish that I could give you a simple rule to guide you as to when this section is necessary and when it isn't, but there is no rule. You have to make this decision based on your instinct regarding the impact on implementation of omitting this section.

Section 4 causes the most discussion on the tentative job description. A company that takes internal politics very seriously never allows this section to be

to me (the DP manager) what jobs my people are going to be working on?" Yes.
included. The reaction is, "You mean that the project manager is going to dictate
That's the intent. We don't want DP to be enhancing an accounts payable system with a new report instead of working on the project, which has a known payback.

What can happen is that a quick, easy, one-shot job may come up that really does have a good return on investment. It looks like the Project Plan should be interrupted to do this new thing. However, part of the decision process should be the understanding of how much project slippage will be incurred and how much of the project's potential benefits will be delayed, as compared to the benefits of this one undertaking. A good company will allow you to include Section 4. A company with a lot of internal politics will never allow this section to be included.

In 90% of the job descriptions, Section 5 is included. The possible reasons for deviation are:

1. The company wants you to come back and conduct the replan. (That's OK if you are paid.)
2. The time period for the replan (nine months) is changed to some other quantity (not to exceed 12 months).

Section 6 is either included in its entirety in the project manager's job description, or it is not included at all. There are only two acceptable reasons for not including it:

1. The personnel department has a training coordinator in place who assumes these responsibilities.
2. A separate training coordinator is appointed because this education area appears to be a full-time job.

Work your way through the sections of Figure 11.1, making changes as necessary. The final draft of the project manager's job description should be reviewed with your executive sponsor. He or she can use your material to put this job position in place right now, even though you have not yet completed your study.

HOW TO SET UP PROCEDURES FOR TASK TRACKING AND PROJECTION

Task tracking can be accomplished with a series of five forms, B1 through B5, which are than used to develop a single page, the B6 progress report. This report can be reviewed by top management to determine project progress, as well as to monitor projected project completion dates. The six forms are included as Figure 11.2.

Report B1 (Planned Manday Summarization), derived from the Implementation Plan, has two purposes: The period totals are applied to Column A of Report B5; and the cumulative totals are plotted on Report B6 as the Original Manday Plan.

Report B2 (Manpower Availability) is to be submitted weekly to allow tracking of which persons were actually available on which days. This is for tracking purposes only and not for projections.

Report B3 (Daily Task Performance Log) is distributed to each person having responsibility for task completion, and it lists the tasks for which that person is responsible. Each day, the person indicates the hours of activity expended on each task. Note that the hours for a specific day do not have to total eight hours; it is understood that people will also be performing activities that aren't task related, which are not recorded by this procedure. An individual department may wish to modify this policy and use the form to report all activity.

Report B4 (Task Reporting) is the log used to represent the collective detail of the total project. The calculated actual mandays on this report should not be logged into Report B5 until the task has been completed. Do not allow the logging of partial completions.

Report B5 (Actual Manday Summarization) is used to evaluate planned versus actual performance, calculate the variance percentage, and statistically project the probable completion of the project.

Report B6 (Progress Report) is a single-page chart illustrating the status of the project for top management. It displays:

- The original manday plan
- The amount of planned mandays in the original plan that have been expended to date
- The actual mandays that have been expended to date
- A statistical projection of the mandays remaining that are required to complete the project

You have noticed by now that this project management approach is a very simple manual method. The preparation time consumes about four hours per week. It is not the only approach, but does have one big advantage—it works.

In a study, you may find a company that wants to use the DP facility or a personal computer to maintain the data for project management. That's OK—do it. You will perform the same steps as the charts in Figure 11.2 illustrate, only it will be faster.

If you want to utilize the described manual approach, which is usually applicable, Figure 11.3 lists the blank forms that are required.

HOW TO SET UP PROCEDURES FOR TRACKING OF ACTUAL COSTS AND BENEFITS, AND PROJECTION OF FURTHER COSTS AND BENEFITS

The major problem in benefit tracking is the identification of the base numbers to be used for tracking. Figure 11.4 provides a basic format for benefit tracking. You want to be able to clearly identify, for example, that inventory was reduced due to installing material requirements planning, or selling obsolete inventory, or selling excess inventory, or eliminating physical staging. If you do not go through the effort to clearly identify the benefit source, you encounter many political problems in proving that planned benefits actually became a reality.

Benefits should be categorized by benefit impact areas within manufacturing functions. If a benefit has been defined for a benefit impact area, then tasks should be established to measure that benefit. Each benefit should be evaluated quarterly. *Careful consideration must be given to how benefits are measured.*

Cost tracking, on the other hand, is relatively easy to measure, since costs can be related to when a check is prepared for payment. Costs (as shown in Figure 11.5) should be totaled by quarter.

When you define task tracking, you want one page for management to review. You also want a single page outlining the cost/benefit area.

In the justification phase of your study, you summarized all costs and benefits by quarter for a break-even analysis. You will also go through exactly the same steps with actual instead of planned data. The only difference is that instead of preparing a new break-even chart, you should superimpose actual data on the chart of planned data that you prepared in the justification phase. Review Figure 11.6. The solid lines represent the original plan, the broken lines show what is actually happening.

Another chart that you might consider is one depicting only variances of planned to actual costs and benefits. Which charts you decide to use will vary, depending on what is comfortable for the company that you are working with.

You have now concluded the definition of how you will manage the implementation of your study. Preparing the final report and conducting the final presentation are all that remain for you to complete the study.

FIGURES

Job Description

Name: ____________________

Title: Manufacturing Control Systems Project Manager

Supervised by: ____________________

Principal Function

Decrease costs and increase profits through the management of the implementation of manufacturing systems.

General Responsibilities

Management of the scheduling and implementation of Manufacturing Systems as defined by this report.

Specific Responsibilities

1. Act as the communications interface between those persons assigned specific task responsibilities and top management.
   A. Conduct weekly reviews with persons responsible for task completion.
   B. Conduct monthly reviews with top management.
2. Monitor the task implementation plan.
   A. Track task performance.
   B. Statistically project completions based on actual performance.
   C. Resolve all reasons (problems/excuses) for nonperformance of tasks.
3. Monitor the cost and benefit implementation plan.
   A. Track actual cost and benefit performance.
   B. Periodically project statistically the actual breakeven based on the costs and benefits being realized.
   C. Keep management informed on all significant cost/benefit variances to the original plan.
   D. Prepare top management proposals on methods to increase benefits and decrease costs.
4. Control changes to the existing plan.
   A. Catagorize and prioritize all "quick," "easy," "one shot," and "really hot" interruptions to the basic implementation plan.
   B. Adjust the plan to emphasize those projects with a greater profit potential than anticipated, and de-emphasize those with cost overruns.
5. Conduct the replanning of the implementation plan annually. (Note: The first replan session is to be performed nine months after the original.)
6. Be responsible for all user education.
   A. Conduct promotional campaigns to promote education.
   B. Schedule and monitor all in-house education as films, video tapes, speakers, and so forth.
   C. Schedule, monitor, and control all plant expenses on all external education as classes requiring tuition and travel.
   D. Promote, sponsor, and develop attendance in professional societies as APICS and AIIE.

FIGURE 11.1 Sample project manager's job description.

PLANNED MANDAY SUMMATION

REPORT B1

PAGE 3 of 3

DEPT- -PERSON	TIME PERIOD ENDING																TOTAL
	7/6	7/13	7/20	7/27	8/3	8/10	8/17	8/24	8/31	9/7	9/14	9/21	9/28	10/5	10/12	10/19	
DATA PROCESSING																	
- E. BROWN	-0-	-0-	-0-	-0-	-0-	-0-	-0-	5.0	-0-	1.0	-0-	-0-	-0-	-0-	-0-	-0-	23.0
- G. CABB	6.0	-0-	7.0	7.0	4.0	-0-	-0-	5.0	7.0	4.0	-0-	9.0	-0-	-0-	-0-	-0-	104.5
- D. COLD	3.5	-0-	7.0	1.5	-0-	5.0	4.0	-0-	-0-	-0-	-0-	-0-	5.0	-0-	-0-	-0-	82.5
- H. HALD	-0-	-0-	-0-	-0-	-0-	-0-	2.5	2.5	-0-	-0-	10.0	5.0	2.0	5.0	-0-	-0-	36.5
- M. SMITH	3.0	-0-	-0-	3.5	2.5	4.0	4.0	2.0	2.0	-0-	-0-	-0-	-0-	-0-	-0-	-0-	52.0
- J. THOMAS	7.0	-0-	3.5	3.0	-0-	10.0	-0-	-0-	-0-	-0-	-0-	-0-	2.0	0.5	1.5	1.5	95.0
SUBTOTAL	19.5	-0-	17.5	15.0	6.5	19.0	10.5	14.5	9.0	5.0	10.0	14.0	9.0	5.5	1.5	1.5	393.5
USER																	
- J. BLUE	-0-	-0-	-0-	-0-	-0-	-0-	5.0	-0-	-0-	-0-	-0-	-0-	-0-	-0-	-0-	-0-	49.0
- C. HARDY	-0-	3.5	-0-	-0-	2.5	-0-	-0-	5.0	-0-	-0-	-0-	-0-	-0-	-0-	-0-	-0-	11.0
SUBTOTAL	-0-	3.5	-0-	-0-	2.5	-0-	5.0	5.0	-0-	-0-	-0-	-0-	-0-	-0-	-0-	-0-	60.0
PERIOD TOTAL	19.5	3.5	17.5	15.0	9.0	19.0	15.5	19.5	9.0	5.0	10.0	14.0	9.0	5.5	1.5	1.5	454
CUMULATIVE	299	303	320	335	344	363	379	398	407	412	422	436	445	451	452	454	

FIGURE 11.2 Completed examples of task tracking charts.

MANPOWER AVAILABILITY

REPORT B2

DEPT. DATA PROCESSING

PAGE 1 of 5

HOURS AVAILABLE FOR TASK PERFORMANCE

PERSON	FEB.			MARCH																						APRIL									
DAY	26	27	28	1	2	5	6	7	8	9	12	13	14	15	16	19	20	21	22	23	26	27	28	29	30	2	3	4	5	6	9	10	11	12	13
E. BROWN	8	8	8	8	8	8	8	8	8	8																									
G. CABB	8	8	8	8	8	V	V	V	V	V																									
D. COLD	8	8	8	8	8	8	8	8	8	8																									
H. HALD	10	11	8	10	10	9	9	8	8	8																									
M. SMITH	8	8	8	8	S	S	8	8	8	S																									
J. THOMAS	8	10	8	V	V	8	8	8	V	V																									

FIGURE 11.2 *(continued)*

DAILY TASK PERFORMANCE LOG

REPORT B3

DEPT. DATA PROCESSING
PERSON G. CABB

PAGE 1 of 4

TASK NUMBER	DATE TASK ACTUALLY		MONTH																					TOTAL HOURS	TOTAL DAYS
	STARTED	COMPLETED	FEB.			MARCH																			
			DAY																						
			26	27	28	1	2	5	6	7	8	9	12	13	14	15	16	19	20	21	22	23	26	27		
9.7	2/26	2/26	2																						2.0	0.5
9.6	2/27	2/28		8	4																				12.0	1.5
9.1	3/2	3/5					8	2																	10.0	1.5
9.2	3/8	3/14									3	5			1										9.0	1.0
9.3	3/13	3/19												3	0.5	3.5	3	2							12.0	1.5
9.5	3/5	3/14						6	1	3					1										11.0	1.5
4.2	3/14	3/20													2.5	4	3	4	2						15.5	2.0
3.7	3/20																		3	7	8					
1.3																										
1.4																										
1.28																										
1.13																										
1.14																										
1.15																										
1.18																										
1.19																										
1.20																										
1.22																										
1.23																										
1.25																										
1.8																										
1.30																										

FIGURE 11.2 *(continued)*

TASK REPORTING

REPORT B4

DEPT. DATA PROCESSING
PERSON G. CABB

PAGE 1 of 2

TASK NUMBER	DATES				TASK ACTUALLY PERFORMED BY (IF DIFFERENT)	TIME					DATE
	START		FINISH			DURATION		MANDAYS			
	PLAN	ACTUAL	PLAN	ACTUAL		PLANNED DURATION IN DAYS	ACTUAL TOTAL DAYS	PLANNED	ACTUAL TOTAL HOURS	ACTUAL MAN DAYS	LOGGED IN RPT. B5
9.7	2/20	2/26	2/22	2/26		2.0	1.0	1.0	2.0	0.5	3/22
9.6	2/27	2/27	3/3	2/28		2.0	2.0	1.0	12.0	1.5	3/22
9.1	3/4	3/2	3/5	3/5		1.0	2.0	1.0	10.0	1.5	3/22
9.2	3/6	3/8	3/7	3/14		1.0	5.0	1.0	9.0	1.0	3/22
9.3	3/8	3/13	3/9	3/19		1.0	5.0	1.0	12.0	1.5	3/22
9.5	3/12	3/5	3/13	3/14		1.0	8.0	0.5	11.0	1.5	3/22
4.2	3/14	3/14	3/23	3/20		7.0	5.0	7.0	15.5	2.0	3/22
3.7	4/4	3/20	4/11			5.0		5.0			
1.3	4/12		4/19			5.0		5.0			
1.4	4/20		4/23			1.0		1.0			
1.28	4/24		5/5			7.0		7.0			
1.13	5/6		5/8			2.0		2.0			
1.14	5/9		5/10			1.0		1.0			
1.15	5/11		5/22			7.0		7.0			
1.18	5/23		5/28			3.0		3.0			
1.19	5/29		6/1			3.0		3.0			
1.20	6/4		6/13			7.0		7.0			
1.22	6/14		6/18			2.0		2.0			
1.23	6/19		6/27			6.0		6.0			
1.25	6/28		7/9			7.0		7.0			
1.8	7/10		7/19			7.0		7.0			
1.30	7/20		7/24			3.0		3.0			

FIGURE 11.2 *(continued)*

ACTUAL MANDAY SUMMARIZATION

REPORT B5

PAGE 1 of 1

DEPARTMENT----- -- PERSON	A TOTAL DEPT/PERSON MANDAYS (FROM RPT B1)	B PLANNED MANDAYS THAT WERE EXPENDED (FROM RPT B4)	C ACTUAL MANDAYS THAT WERE EXPENDED (FROM RPT B4)	D PLANNED VS. ACTUAL VARIANCE PERCENT (C÷B=D)	E PLANNED MANDAYS REMAINING (A-B=E)	F PROBABLE MANDAYS REMAINING (DxE=F)
DATA PROCESSING						
- E. BROWN	23.0	-0-	1.0	N/A	23.0	23.0
- G. CABB	104.5	17.5	12.5	.71	87.0	62.0
- D. COLD	82.5	23.0	14.5	.63	59.5	37.5
- H. HALD	36.5	8.5	3.0	.35	28.0	10.0
- M. SMITH	52.0	21.0	13.5	.64	31.0	20.0
- J. THOMAS	95.0	15.0	13.0	.87	80.0	69.5
SUBTOTAL	393.5	85.0	57.5		308.5	222.0
USER						
- J. BLUE	49.0	14.0	6.0	.43	35.0	15.0
- C. HARDY	11.0	-0-	-0-	N/A	11.0	11.0
SUBTOTAL	60.0	14.0	6.0		46.0	26.0
TOTALS	453.5	99.0	63.5	X	354.5	248.0

FIGURE 11.2 *(continued)*

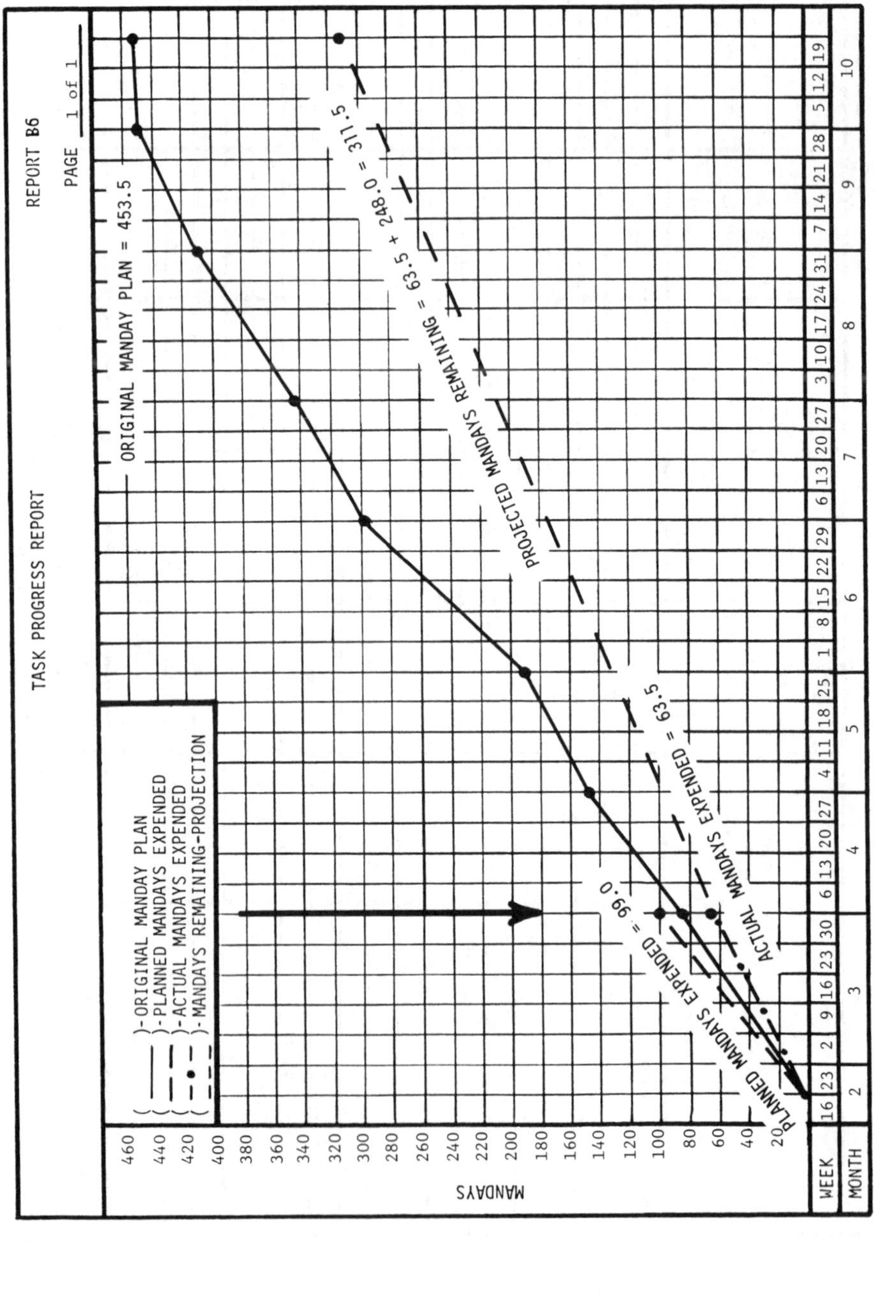

TASK PROGRESS REPORT

REPORT B6

PAGE 1 of 1

(———)-ORIGINAL MANDAY PLAN
(— —)-PLANNED MANDAYS EXPENDED
(—•—)-ACTUAL MANDAYS EXPENDED
(- - -)-MANDAYS REMAINING-PROJECTION

ORIGINAL MANDAY PLAN = 453.5

PLANNED MANDAYS EXPENDED = 99.0

ACTUAL MANDAYS EXPENDED = 63.5

PROJECTED MANDAYS REMAINING = 63.5 + 248.0 = 311.5

MANDAYS

460 440 420 400 380 360 340 320 300 280 260 240 220 200 180 160 140 120 100 80 60 40 20

WEEK 16 23 2 9 16 23 30 6 13 20 27 4 11 18 25 1 8 15 22 29 6 13 20 27 3 10 17 24 31 7 14 21 28 5 12 19

MONTH 2 3 4 5 6 7 8 9 10

FIGURE 11.2 *(continued)*

PLANNED MANDAY SUMMATION																	REPORT B1 PAGE______
DEPT- -PERSON	TIME PERIOD ENDING																TOTAL
PERIOD TOTAL																	
CUMULATIVE																	

FIGURE 11.3 Blank task tracking charts.

MANPOWER AVAILABILITY

REPORT B2

DEPT.__________

PAGE________

PERSON	HOURS AVAILABLE FOR TASK PERFORMANCE
	MONTH
	DAY

FIGURE 11.3 *(continued)*

DAILY TASK PERFORMANCE LOG

REPORT B3

DEPT. ______________

PERSON ______________

PAGE ________

TASK NUMBER	DATE TASK ACTUALLY		MONTH / DAY	TOTAL HOURS	TOTAL DAYS
	STARTED	COMPLETED			

FIGURE 11.3 *(continued)*

TASK REPORTING

REPORT B4

DEPT. ____________

PERSON ____________

PAGE ________

TASK NUMBER	DATES				TASK	TIME					DATE
	START		FINISH		ACTUALLY	DURATION		MANDAYS			
	PLAN	ACTUAL	PLAN	ACTUAL	PERFORMED BY (IF DIFFERENT)	PLANNED DURATION IN DAYS	ACTUAL TOTAL DAYS	PLANNED	ACTUAL TOTAL HOURS	ACTUAL MAN DAYS	LOGGED IN RPT. B5

FIGURE 11.3 *(continued)*

ACTUAL MANDAY SUMMARIZATION						REPORT B5 PAGE ______
DEPARTMENT----- -- PERSON	A TOTAL DEPT/PERSON MANDAYS (FROM RPT B1)	B PLANNED MANDAYS THAT WERE EXPENDED (FROM RPT B4)	C ACTUAL MANDAYS THAT WERE EXPENDED (FROM RPT B4)	D PLANNED VS. ACTUAL VARIANCE PERCENT (C÷B=D)	E PLANNED MANDAYS REMAINING (A-B=E)	F PROBABLE MANDAYS REMAINING (DxE=F)
TOTALS				X		

FIGURE 11.3 *(continued)*

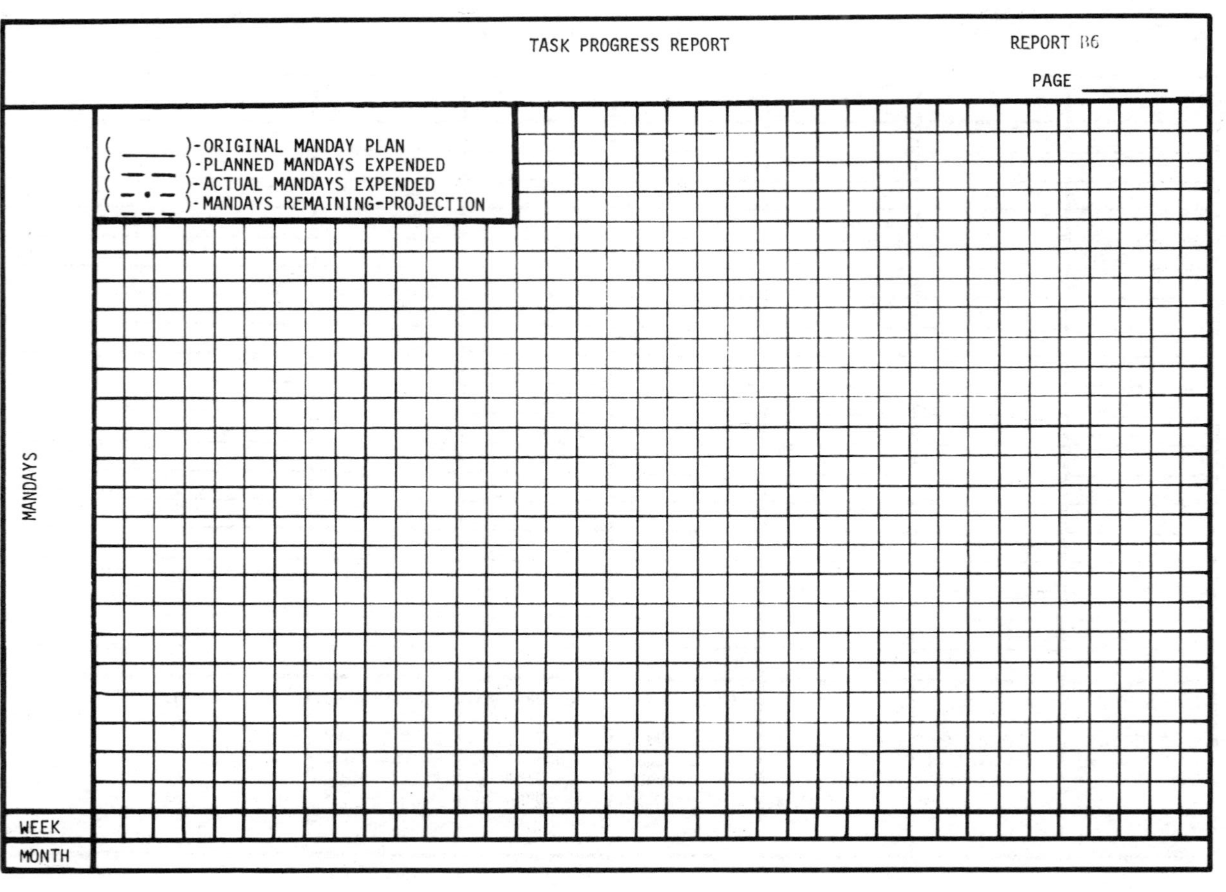

FIGURE 11.3 *(continued)*

	Measurement		Initial Status		Planned		Actual		Variance	
Functional Area	Area	Technique, Tool, or Procedure	At Date	Amount	By Quarter	Amount	On Quarter	Amount	Amount	Percentage

FIGURE 11.4 Sample benefit measurement format.

	Planned		Actual Transactions				Variance	
Cost Area	By Quarter	Amount	Item Number	Description	Date	Amount	Amount	Percent

FIGURE 11.5 Sample cost measurement format.

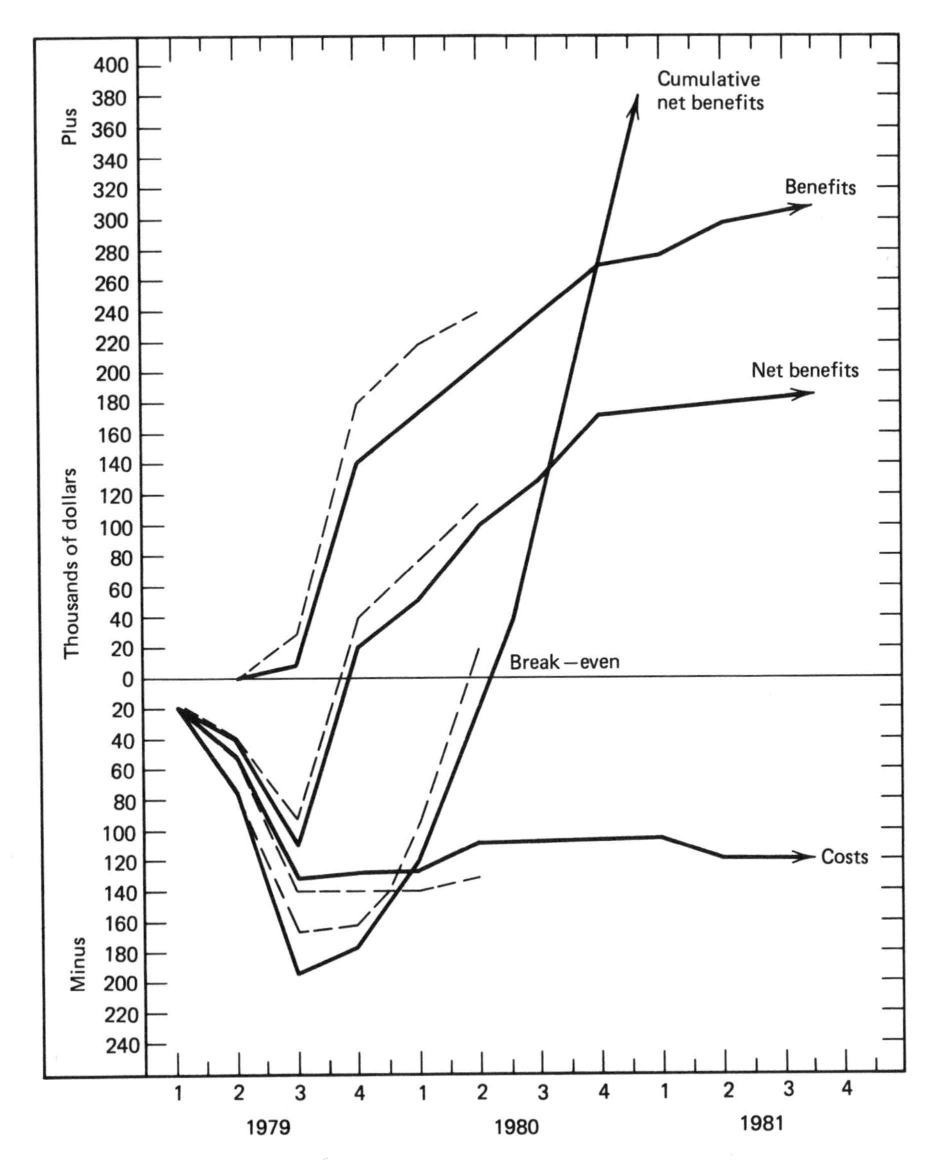

FIGURE 11.6 Sample break-even chart comparing planned to actual data.

CHAPTER TWELVE

PREPARING THE FINAL REPORT AND FINAL PRESENTATION

PREPARING THE FINAL REPORT

At this point, allow me to make some assumptions. You followed the suggestions made in earlier chapters about framing your final report up front. Since you submitted flip chart pages to typing, you included these in your final report. As typing was completed, you edited the material to make it as precise as possible.

If my above assumptions are not true and you did not do these things during your study, you should do them now. Keep in mind that your study company is going to evaluate your productivity—formally or informally. If you now waste a week on a mechanical function like final report preparation, your productivity will be low. Final report structuring and basic editing should be done, by you, during the study.

You and your team have completed the main study. You should have a draft of the study report. *Only have one.* If your typist has a word processor or a magnetic tape/card machine, that is great, but only have one paper copy available. To do a final edit, you give the entire document to the first team member. He or she reads page 1 and passes it to the second team member. The pages circulate through the entire team, each person making corrections on the one set of draft pages. Once the edit is complete, the document can now be submitted to typing for final copy preparation.

Here are some other considerations for the preparation of your final report:

How many copies will you need? Who (by name) will receive them?
Do you wish to categorize your distribution as to:
- Initial distribution only for information purposes
- Distribution to be updated on some periodic basis
- Partial final report distribution, such as only the implementation planning pages applicable to the receiver.

What type of binders will you use?
- Loose leaf, 3 ring (most effective)
- Spiral bound (difficult to use)
- Hard bound (too expensive)

Do you want the company logo on the cover of the binder? What will that cost? What lead time is necessary?
Will a commercial printer run off the copies for you, or will the typist do it? What lead time is necessary?
Are index tabs going to be used? Who makes them? (Index tabs are a good idea.)
Who will assemble the final report (pages, binder, tabs, etc.)?
How will distribution be accomplished? (After the final presentation is a good idea.)

I'm sure that you've noticed a couple of references to "lead time" in the foregoing considerations. I usually discuss these considerations with the team during the second week of a six-week study so that I do not get into a lead time problem. However, I once had a company who wanted a three-color logo embossed on the final report binders. The lead time was 10 weeks. I talked them into something less fancy so that our report would be available when the Final Presentation was completed. It's great to want to illustrate the final report as a class product, but you also have to be practical.

PREPARING THE FINAL EXECUTIVE PRESENTATION

The first thing that you and your team should decide on is who you are going to present it to. This is a key factor for the determination of the content of your presentation. Identify by name and title those that you plan to have in attendance. Then check out the schedules of each of these persons and derive a presentation date that is free of schedule conflicts. This presentation date should be as close to the study completion date as possible (given time for final report preparation), because that is when the project enthusiasm is the highest.

Now on a flip chart page, list the names of your intended audience down the left side of the page. Across the top of the page, list key topics that you plan to cover in your presentation. Put an "X" by the name of each person who will have key interest in a specific topic. Now evaluate your chart. Do you have a topic that only one or two people (out of 15) will be interested in? Are there other topics that should be included? Have you covered the "hot buttons" of the key executives? Adjust your chart to reflect what you are going to cover in the presentation.

To build your presentation, divide a blank flip chart page into nine equal squares. Number them 1 to 9. Title square 1 "Cover Page." Title square 9 "Approval Page." Title square 2 "Telling you what I'm going to tell you." Title square 8 "Telling you what I told you." You have squares 3, 4, 5, 6, and 7 for the body of your presentation. Which is the hottest topic? Put it in square 3. Which is the second hottest topic? Put it in square 7, since you want to balance the interest level throughout your presentation.

You now have a basic frame for your presentation. Decide which team members are going to present which parts of the total pitch. When, for example, your team has decided that John Smith is best qualified to talk about Inventory Management, ask John to list the key points that he plans to cover.

When everyone who is participating in the presentation has framed his or her own segments, do a dry run. Have each person stand up in the proper presentation sequence and discuss what he or she plans to cover. How did it flow? Were areas redundant? Was something missing? Repeat the presentation until it appears to flow smoothly.

Each presenter should now further develop his individual foil pages. Do not type them. Make foils of the handwritten pages. Again do a dry run. Allow the team to make suggestions as to where points should be added, stressed, or deleted. Do a refinement and have all of the material typed, using a large font that can be easily seen on an overhead projector screen.

You may be thinking that this looks like a lot of work to just make a pitch. Yes, it is a lot of work, but it's not just a pitch. It's the pitch that will make your study live or die. If you do several final presentations wrong, you soon won't have to worry about doing any at all. Do them right and your reputation will keep your phone ringing with new business.

Okay, now try the next dry run. Select a training audience for your study team to pitch to. The best people are those who are directly responsible to the key executives that you plan to have in your final presentation. You only need about four or five people. Give them the presentation as if it was the real thing. Get their comments. What was good? What was bad? Make adjustments to the presentation.

Now you should have the basics for a clean, effective presentation. Rehearse it again. By this time, your team is saying, "Joe, don't put your hands in your pockets." Or "I think a pointer would be useful to explain this area." You are building a comfort level in your team. How many times should you do this final "dress rehearsal"? Normally once, but it varies. I have done as many as six final dry runs. And now your pitch is ready, your final report is ready, and the end is in sight.

CONDUCTING THE FINAL EXECUTIVE PRESENTATION

You are ready. Your team is ready. Your Final Study Reports are stacked up in the back of the room to hand out when the presentation is over (you don't want people leafing through the report while the presentation is being given).

All of the considerations that were covered under Interim Reviews are applicable here. You've done your homework. Just relax in the back of the room and let your team give the final presentation. Allow for questions and answers several times during the presentation as well as at the conclusion.

You (and your team) want only one thing out of this presentation. Remember when we were laying out the pitch, I said label square 9 as an approval page. This page directs certain questions to top management:

1. Do you agree?
2. Can we start the implementation?
3. What is the start date?

This is called asking for the order. It will not be given to anyone who does not ask. You must ask, "Mr. President, since you agree with our definition of requirements and solutions, would you please give us a start date for implementation."

Now you have a commitment. Now you have a study that will be converted to reality. Now you have a reference for potential future studies.

At this point, you've done it. The study is over. Go out to dinner with your study team. Turn in your preliminary expense accounts.

And of course, good luck on your next study.

APPENDIXES

APPENDIX A

SAMPLE NOMINATION FORM

The following Nomination Form is designed for the evaluation of a potential study. You must tailor it to suit your specific needs. For example, if you are a private consultant, you should omit all of "Section O—Branch Office Control Information," because it would not be applicable to your operation.

Note that some of the questions are very direct. Some may be embarrassing to answer. This is done on purpose. There are two major points that you must determine using this form.

1. How serious is the company about doing a study? You learn to avoid the folks who are very nice, will listen to you when you spend the money (your money) to fly in and present to them, but will never do a study.
2. Is the Company willing to expose itself to you? Will you be told the real situation? You will be able to tell if questions are left blank or have vague answers. The whole study process is a two-way exchange. If the company representatives appear to be hesitant about leveling with you regarding the company's status, you'll have a problem in trying to help them.

If you are in a situation where you work for a company that has branch offices, you'll want to compose a cover letter to this Nomination Form that explains the benefits of your program to all of the sales representatives. Then you'll have a field force submitting nominations to you for evaluation.

If you are a private consultant (working for yourself), consider advertising your services in publications such as the APICS (American Production and

Inventory Control Society) monthly magazine. When you get a nibble, send the Nomination Form.

If you work for the company for which you want to do the study, it is still a worthwhile excercise to complete the Nomination Form. All too often, things are assumed on an internally performed study. The Nomination Form tends to reduce the assumptions (which are usually wrong) to facts.

I have said it before and let me say it again. *Do not* let greed or hunger influence your decision to do a study. If you do, you should be independently wealthy before you start (and you'll be broke when you finish). You should rank your nominations in terms of potential success. Do the guaranteed winners first. If you don't have "guaranteed winners," or even "maybe winners," and all you have is probable losers, why do them? Walk away. Say "Thank you, but no thank you."

As you can tell, the Nomination Form is critical. It allows you to evaluate potential winners. Tailor it to your needs very carefully.

Nomination
for a
Manufacturing Information System Study

Submit Nomination to:

Mr. John Doe
1000 Every Avenue
Our Town, State Zip

Phone: *666/999/1234* for questions

Submission Date: ______________________________

Company Name: ______________________________

Desired Study Start Date: ______________________________

Alternative Study Start Date: ______________________________

Desired Study Site: ______________________________

Alternate Study Site: ______________________________

SECTION O
BRANCH OFFICE CONTROL INFORMATION

1. *General Data*
   A. Nomination submitted by: ______________________
      Position: ______________________
   B. Branch Office:
      Name: ______________________
      Number: ____________ Region: ____________
      Mailing Address: ______________________
      ____
      Physical Address: ______________________
      ____

2. *Branch Personnel*
   A. Management

	Name	Phone Internal	Phone External
Branch Manager			
Marketing Manager			
Systems Engineering Manager			

   B. Support Personnel on the Study

Name	Title	Phone Internal	Phone External

3. *Potential Sales Evaluation*
   A. Hardware

Product Number	Product Description	Quantity	Extended Value	Projected Book Date	Projected Install Date
			$	/ /	/ /
			$	/ /	/ /
			$	/ /	/ /
			$	/ /	/ /
			$	/ /	/ /
			$	/ /	/ /
			$	/ /	/ /
		Subtotal	$		

B. Software

Product Number	Product Description	Quantity	Extended Value	Projected Install Date
______	______	______	$ ______	/ /
______	______	______	$ ______	/ /
______	______	______	$ ______	/ /
______	______	______	$ ______	/ /
______	______	______	$ ______	/ /
______	______	______	$ ______	/ /
______	______	______	$ ______	/ /
______	______	______	$ ______	/ /
		Subtotal	$ ______	
		Gross total	$ ______	

4. *Branch Commitment*

I have reviewed the information in this nomination and agree that the Branch will support this study with the ______(Quantity) individuals that were indicated at ______% of their time. The identified potential sales evaluation is a true approximation of the business case that will result from the study. This business will become part of this Branch's Marketing Plan, if the study is approved and is successful.

Type name ______ Sign name ______

______ ______

System Engineering Manager Date

______ ______

Marketing Manager Date

______ ______

Branch Manager Date

SECTION I
GENERAL SUBMISSION INFORMATION

1. *Company Identification* (where study will be performed)
   A. Company: ______________________________
   B. Mailing address: ______________________________
   ____
   ____
   C. Physical Address: ______________________________
   ____
   ____
   D. Phone: ____/____/__________
2. *Company Ownership*
   A. This Company is owned by: ______________________________
   B. The above Company is owned by: ______________________________
   C. The above Company is owned by: ______________________________
   D. The above Company is owned by: ______________________________
   E. Study approval will be required by the following company: ______
   ____
3. *Submitter Identification*
   A. Your Name: ______________________________
   B. Your Title: ______________________________
   C. Your Responsibilities: ______________________________
   ____
   ____
   D. Number of months that you have been employed by this company: ________ Months
4. *Key Reasons for doing a Study*
   A. Identify the key reasons why you think that a study should be done at this company. Sequence your reasons by most important reason first.
   ____
   ____
   ____
   ____
   ____
   ____
   ____
   ____
   ____
   ____
   ____

B. Identify the managers in your company that support the above identified reasons.

Name	Title

5. *Key Company Management*
(Please, also attach an organization chart)

Name	Title

If the executive sponsor of this study is known at this time, indicate with an "E.S." after the name in the above Key Management List.

SECTION II
NOMINATION PURPOSE

1. *Awareness of Methodology*
   - A. The study methodology, purpose and intended result is understood by:
     Me, the submitter.
     ______ Fully
     ______ Somewhat—I need more information
     ______ Not at all—I need a complete explanation of why this study should be done
     The top management of the company.
     ______ Fully
     ______ Somewhat—Some managers need more information
     ______ Not at all—A complete explanation of why this study should be done needs to be provided.

2. *Proposed Study Type*
   - A. This nomination is submitted with the intent of conducting:
     ______ A complete study which would include:
     Problem definition
     General system design
     Implementation planning
     Cost/benefit justification
     ______ An Implementation Study (implies that you absolutely understand your problem areas and you have already justified the implementation of a solution) which would include:
     General system design
     Implementation planning

3. *Commitment*
   - A. Three to five full time team members will be required for six to eight weeks on a Complete Study or three to five weeks for an Implementation Study. The team members must be key management people from areas as Material Control, Engineering, Manufacturing, and Data Processing. Has this company expressed willingness to make this commitment?
     ______ Yes
     ______ No
   - B. A conference room will be required that should be twice the size of what would normally be required for the number of people on the study team. Will such a facility be 100% available for the complete duration of the study?
     ______ Yes
     ______ No

4. *Scope*

   A. The scope of this study is to include the following plants and locations:

Plant/Location Name	City/State	Percent of Contribution to study

SECTION III
GENERAL COMPANY DATA

1. The major products of this company are: ______________________

____

____

2. This company performs the following type of manufacturing.

   ___% Major project oriented
   ___% Pure make-to-order job shop
   ___% Make to stock and assemble to order
   ___% Make to stock
   ___% Other: ______________________
   100% Total

3. What is the origin of this company? How did it start? When? What significant events occurred from its start to now? ______________________

____

____

____

4. What trends/changes are anticipated for this company during the next 10 years? Consider the impact of techniques such as the Japanese "Just-in-Time" concept.

____

____

5. Sales

   A. Total Sales Volume

	Study company	Top owning company
Last year	$ ______ million	$ ______ million
This year (est)	$ ______ million	$ ______ million
Next year (est)	$ ______ million	$ ______ million

   B. Total end items sold are (by type): ______________________

   C. Acceptable profit margin (top management comfort level)

   Percent of end items that are sold below an acceptable profit margin ______%

   Percent of end items that are sold at an acceptable profit margin ______%

   Percent of end items that are sold above the acceptable profit margin ______%

6. Data Processing Installation (Hardware/Software)

A. Hardware

	Vendor	Type	Qty.
Central processing unit	______	______	______
Disk	______	______	______
Tape	______	______	______
Printer	______	______	______
Terminals	______	______	______

B. Software

	Vendor	Type	Qty.
Main operating system	______	______	______
Data base system	______	______	______
Data communication system	______	______	______

SECTION IV
APPLICATION (FUNCTION) STATUS

	The Present Percent Performed Is		Desired Study Emphasis			
Area	Manually	Via DP	High	Medium	Low	None
Business Planning	____%	____%	____	____	____	____
Forecasting	____%	____%	____	____	____	____
Demand Analysis	____%	____%	____	____	____	____
Production Planning (with resource testing)	____%	____%	____	____	____	____
Master Schedule Planning (with rough cut capacity testing)	____%	____%	____	____	____	____
Customer Order Servicing	____%	____%	____	____	____	____
Inventory Accounting	____%	____%	____	____	____	____
Bill of Material Management	____%	____%	____	____	____	____
Stores Location Management	____%	____%	____	____	____	____
Material Requirements Planning	____%	____%	____	____	____	____
Shop Order Release Control	____%	____%	____	____	____	____
Capacity Requirements Planning	____%	____%	____	____	____	____
Routing Data Control	____%	____%	____	____	____	____
Tool Data Control	____%	____%	____	____	____	____
Facility Data Control	____%	____%	____	____	____	____
Purchasing	____%	____%	____	____	____	____
Receiving	____%	____%	____	____	____	____
Shop Floor Control	____%	____%	____	____	____	____
Product Costing	____%	____%	____	____	____	____
Plant Maintenance	____%	____%	____	____	____	____

APPENDIX B

SAMPLE COMPANY QUESTIONNAIRE

Let us assume that you have ten nominations for studies in your hand. You have evaluated them in terms of which are most likely to be successful. That new list constitutes four out of the original ten. Now, you would like to obtain some in-depth understanding of each of those four companies. You'd like to do this before you spend travel money and find out that a potential winner is a probable loser.

One potential approach to resolving who is a winner is to use a company questionnaire. You make some phone calls to your company contact (person who sent in the nomination form) and explain that you would like some additional information. Describe the questionnaire on the phone. Explain that you would like the questionnaire completed and returned prior to your first visit, so that you will have a better understanding of the company, and can therefore tailor your material to specific company needs.

The questionnaire, like the nomination form, is a sample. I strongly suggest that you tailor the attached questionnaire to your specific needs.

You may notice that the functions in "Section IV—Function Analysis" do not cover all of the functions in "Section IV—Application (Function) Status," of the nomination form. You should likewise adjust the questionnaire to areas of interest to you.

The questionnaire has been structured into these parts:

Cover Page. This allows you to at a glance tell who sent the questionnaire to you and when.

Section I—General Questionnaire Guidelines. These are instructions to the person who will be completing the document.

Section II—General Submission Information. This is the data necessary for you to relate the Questionnaire to the Nomination.

Section III—Company Characteristics. This provides you data in terms of characteristics (not applications or functions), as to where the company is and where the company is headed. If you review the questions in this section, you will notice a definite structure. In every case, if the first of the two possible answers to a question is selected, the company tends to be a job shop. If the second answer is selected the company tends to be a flow or process shop. It is a simple analysis to scan down the list of responses to evaluate your potential study company.

Section IV—Function Analysis. This section is structured to provide you with information about how the company handles manufacturing functions today versus how they would like to change.

Remember, the attached sample questionnaire was one that worked well for me. I tailored it. One that works well for you will require that you tailor it.

Company Questionnaire
for a
Manufacturing Information System Study

Submit Questionnaire to:

Mr. John Doe
1000 Every Avenue
Our Town, State Zip

Phone: *666/999/1234* for questions

Submission Date: ______________________________

Company Name: ______________________________

SECTION I
GENERAL QUESTIONNAIRE GUIDELINES

This questionnaire is being submitted to you to assist me in the identification of those requirements—present and future—that you feel are vital in conducting your business.

Please review each characteristic carefully, as an item of importance to you today may become unnecessary in the future, if you have any intention of changing the way that you run your business.

The results of this questionnaire will assist me in performing a more effective job in supporting your current as well as future requirements.

Please answer each question as accurately as possible.

SECTION II
GENERAL SUBMISSION INFORMATION

1. *Company Identification*

   A. Company name: ______________________________

   B. Mailing address: ______________________________

   ____

   ____

   C. Physical address: ______________________________

   ____

   ____

   D. Phone: ____/____/__________

2. *Submitter Identification*

   A. Control of the completion of this document was performed by:
   Your name: ______________________________
   Your title: ______________________________

   B. Additional personnel that had input to this document were:

Name	Title

SECTION III
COMPANY CHARACTERISTICS

	Estimated Current Status		Probable Future Status	
1. Your marketing emphasis is based on:				
Features and Flexibility	___		___	
Price and Availability		___		___
2. The quantity of products (end items) that you have available to sell are:				
Many	___		___	
Few		___		___
3. The number of customers that you have are:				
Many	___		___	
Few		___		___
4. The volumes of the product that you sell are:				
Low	___		___	
High		___		___
5. Sales "swapping agreements" with your competitors are exercised:				
Seldom	___		___	
Often		___		___
6. The number of plants that you have are:				
Three or less	___		___	
Four or more		___		___
7. Your plant is layed out by:				
The manufacturing process as drills, mills, and so forth	___		___	
The requirements to make a specific product type		___		___
8. The path of your material handling equipment is:				
Variable	___		___	
Fixed		___		___

	Estimated Current Status	Probable Future Status
9. The bulk of your manufacturing equipment tends to be:		
General purpose, for use on many products	____	____
Specialized for specific products	____	____
10. Your ability to define available plant/machine capacity tends to be:		
Difficult	____	____
Easy	____	____
11. The time required for you to increase plant/machine capacities is normally:		
Short	____	____
Long	____	____
12. The failure of one piece of equipment could shut down your:		
Equipment that failed only	____	____
Plant	____	____
13. If you have multiple plants, they have been geographically located based on:		
Minimum transportation costs	____	____
Labor availability and existing unions	____	____
14. Reduced sales demand could easily affect your plant with a:		
Slow down	____	____
Shutdown	____	____
15. Your shop floor personnel tend to be:		
Skilled craftmen	____	____
Trained operators on specialized equipment	____	____
16. Your plant operation tends to be:		
Labor intensive	____	____
Capital intensive	____	____
17. A general strike would affect your plant with a:		
Shutdown	____	____
Slow down	____	____

(Continued)

	Estimated Current Status		Probable Future Status	
18. You consider your product structures, on the average, to be:				
Narrow and deep	___		___	
Broad and shallow		___		___
19. Your interest in product identification is by:				
Serial number and/or part number	___		___	
Lot number and/or batch number		___		___
20. The design changes on your products are considered to be:				
Many	___		___	
Few		___		___
21. Your use of substitute materials is:				
Occasional	___		___	
Often		___		___
22. The main reason for your Bill-of-Material changes is due to:				
Design changes	___		___	
Change based on the source and availability of materials		___		___
23. The number of features that you offer on your products is:				
High	___		___	
Low		___		___
24. The Bill-of-Material identification of by-products or off-grade yields is considered to be:				
Easy (only a few, if any)	___		___	
Difficult (could be hundreds)		___		___
25. Your long range planning for projected required capacities is in the range of:				
One to two years	___		___	
Two to four years		___		___
26. The impact that production planning has on the utilization of available resources is:				
Low to medium	___		___	
Medium to high		___		___

	Estimated Current Status		Probable Future Status	
27. Your shop schedule is created as a direct output of production planning:				
Seldom	___		___	
Normally		___		___
28. Your build cycle has a logical relationship to the different products that you manufacture:				
Seldom (any product can be followed by any other)	___		___	
Usually (Product B is normally started up after Product Line A shuts down)		___		___
29. Do you ever intentionally shut down the plant for maintenance?				
No	___		___	
Yes		___		___
30. Are your shifts offset (such as 8:00 a.m. to 4:00 p.m. and 8:00 p.m. to 4:00 a.m.) to allow maintenance to be scheduled between shifts?				
No	___		___	
Yes		___		___
31. The planning systems that drive your plant are mainly concerned with:				
Material	___		___	
Capacity		___		___
32. Raw material is identified at the production planning level:				
Seldom	___		___	
Often		___		___
33. In planning for the utilization of available capacity, you find that it is usually:				
Difficult	___		___	
Easy		___		___

(Continued)

	Estimated Current Status		Probable Future Status	
34. Your usage of aggregated production plans is:				
None to medium	____		____	
Medium to high		____		____
35. Your usage of material requirements planning is:				
Daily to weekly	____		____	
Less frequent than weekly		____		____
36. You feel that the quantity of your receiving/receiving inspection inventory is:				
High (more than you would like)	____		____	
Low (less than your competitors)		____		____
37. You feel that the quantity of your work in process inventory is:				
High	____		____	
Low		____		____
38. The frequency that you warehouse work in process inventory is:				
Often	____		____	
Seldom		____		____
39. Intermediate component sales demand for components that go into your products (excluding service parts) is:				
Low	____		____	
High		____		____
40. The quantity of below specification product that you produce is:				
Little to none	____		____	
Some to a lot		____		____
41. You dispose of rejected (below specification) product by:				
Rework or scrap	____		____	
Reprocessing it or selling it as is		____		____
42. The quantity of raw material quality variations that you will accept are:				
Few	____		____	
Many		____		____

	Estimated Current Status		Probable Future Status	
43. Raw material (from supplier) that is not up to standard specifications upon receipt is usually:				
Rejected or reworked	___		___	
Accepted and used		___		___
44. The volume of product that you produce (not the quantity of item types) is considered to be:				
Low	___		___	
High		___		___
45. You consider the number of items that you have in inventory that have a shelf life to be:				
Low	___		___	
High		___		___
46. Your lot sizes are determined by:				
Costs	___		___	
Facility constraints		___		___
47. The sales dollar value of your product per unit is considered to be:				
High	___		___	
Low		___		___
48. When you have a late receipt of purchased material, it is likely to:				
Delay a customer order	___		___	
Shut down the plant		___		___
49. The material planning that you perform with your suppliers is based on:				
The immediate future demands	___		___	
Long term (blanket) contracts		___		___
50. In the selection of your suppliers, you go out for bid and juggle suppliers:				
Often	___		___	
Seldom		___		___

(Continued)

	Estimated Current Status		Probable Future Status	
51. Do you accept material/components from suppliers with some over/under quantity percentage as opposed to the exact purchase order quantity?				
Yes	___		___	
No		___		___
52. Do you expect defect-free material/components from your suppliers and therefore you do not do receiving inspection:				
No	___		___	
Yes		___		___
53. Do you expect (require) your suppliers to ship:				
Infrequently in large lots to obtain lot discounts	___		___	
Frequently (daily) in very small quantities to decrease inventories		___		___
54. Do you expect your suppliers to utilize "just-in-time/total quality control" concepts that are used by the Japanese?				
No	___		___	
Yes		___		___
55. Do you do forecasting?				
No	___		___	
Yes		___		___
56. If you forecast, do you have a safety stock for forecast error which is adjusted as the forecast error varies?				
No	___		___	
Yes		___		___
57. If you forecast, do you have a safety stock to accommodate for a machine not being available as needed (due to maintenance or another product being run)?				
No	___		___	
Yes		___		___

	Estimated Current Status		Probable Future Status	
58. If you forecast, do you have a safety stock to buffer variations in product yield?				
No	___		___	
Yes		___		___
59. Do you adjust (recalculate) safety stocks on a dynamic basis based on current demand as opposed to some fixed quantity?				
No	___		___	
Yes		___		___
60. You have found that the easiest way to produce your product is to?				
Provide shop orders with priorities	___		___	
Provide a schedule of product quantity required each day		___		___
61. Your ideal direct labor work schedule is:				
One shift per day/five days per week	___		___	
Three shifts per day/seven days per week		___		___
62. At any point in time, the number of different products that you have in the process of being manufactured is:				
Large	___		___	
Small		___		___
63. Your product production is measured in:				
Units (piece count)	___		___	
Pounds/gallons/and so on (volume count)		___		___
64. Do you use lot sizes between work stations instead of a machine operator handing one, or a few pieces, to the operator at the next station?				
No	___		___	
Yes		___		___

(Continued)

	Estimated Current Status	Probable Future Status
65. You manufacture your products by utilizing:		
A routing that depicts the steps to be performed by several or many general purpose machines	___	___
A flow line where all or many of the machines are physically in the sequence required to produce the product	___	___
66. The emphasis of the management in your company is to:		
Decrease the inventory	___	___
Balance the line	___	___
67. You measure defects in parts per:		
Hundred	___	___
Million	___	___
68. You fell that quality should be the responsibility of:		
Quality control	___	___
Production control	___	___
69. Your finished goods inventory is owned by:		
Manufacturing	___	___
Marketing	___	___
70. On a yearly basis, the cost of your equipment (as opposed to the cost of your labor) tends to be:		
Low	___	___
High	___	___
71. Your traffic department manages inbound as well as outbound transportation costs:		
No	___	___
Yes	___	___
72. Transportation costs in your company, due to the frequency of inbound deliveries, tend to be:		
Low (large quantity per delivery)	___	___
High (small quantity per delivery)	___	___

	Estimated Current Status		Probable Future Status	
73. The variations that you experience in raw material costs are:				
Few	___		___	
Many		___		___
74. Your concern on the capital budget (fixed assets) is of:				
Low to medium concern	___		___	
Medium to high concern		___		___
75. Your concern on energy costs is:				
Low to medium	___		___	
Medium to high		___		___
76. Your concern on waste disposal costs is:				
Low to medium	___		___	
Medium to high		___		___
77. The number of times that you desire an optimization to reduce input-blend costs on raw materials is:				
Seldom	___		___	
Often		___		___

SECTION IV
FUNCTION ANALYSIS

	Currently		Desired in the Future	
	Yes	No	Yes	No
A. *Business Planning*				
1. Are sales trends analyzed by product?	___	___	___	___
2. Are sales trends analyzed by customer?	___	___	___	___
3. Is an analysis made as to which product is the major contributor to profit?	___	___	___	___
4. Is an analysis made as to the proper level for the evaluation of product trends?	___	___	___	___
5. Are projections realistic and not based on "wish lists"?	___	___	___	___
6. Is customer feedback used to develop projections?	___	___	___	___
7. Are forecasting models used to assist business planning?	___	___	___	___
B. *Forecasting*				
8. Are products grouped into sales families with similar sales trends?	___	___	___	___
9. If applicable, could forecasts be developed for multiple distribution points and consolidated to a single-product forecast?	___	___	___	___
10. Are sales forecasts tracked by comparing actual demand against the forecast?	___	___	___	___
11. Does the sales forecast include an estimate of forecast error?	___	___	___	___
12. Are sales forecasts reviewed regularly by both marketing and production?	___	___	___	___
13. Is the best judgement of the group exercised in improving forecasting data, methods, and techniques used?	___	___	___	___

	Currently		Desired in the Future	
	Yes	No	Yes	No
14. Are changes to the forecast promptly reflected in follow on systems?	___	___	___	___
15. Is an ABC analysis run on profit for each product to determine the product's contribution to the total profit picture?	___	___	___	___
C. *Demand Analysis*				
16. Is demand evaluated in terms of both satisfied and unsatisfied demand?	___	___	___	___
17. Can forecasted sales families be exploded to production planning or master schedule planning items?	___	___	___	___
18. Can models be exploded to options using correct percentages of mix?	___	___	___	___
19. Can multiple forecasts for an item be consolidated?	___	___	___	___
20. Can service part demand be analyzed separate from the dependent demand for the parts?	___	___	___	___
21. Can actual sales be blended to the forecasts?	___	___	___	___
22. Is more than one blend technique available for use?	___	___	___	___
23. Can items be identified for production planning and/or master schedule planning?	___	___	___	___
D. *Production Planning*				
24. Are item net demands aggregated by period to product family net demands?	___	___	___	___
25. Can product family net demands be realistically balanced against management's desired family production plans?	___	___	___	___

(Continued)

	Currently		Desired in the Future	
	Yes	No	Yes	No
26. Are family production plans tested with resource profiles of total machine hours, labor hours, or something else?	____	____	____	____
27. Is the disaggregation of family production plans based on the same period percentages of mix that were used in the aggregation?	____	____	____	____
28. Does a centralized production planning function cover all plants for capacity balance?	____	____	____	____
29. Are production plans current?	____	____	____	____
E. *Master Schedule Planning*				
30. Does the master schedule show what items are to be produced where, when, and in what quantity?	____	____	____	____
31. Is the master schedule realistic and within the capability of the plant to produce?	____	____	____	____
32. Does the master schedule cover the full horizon of the manufacturing and purchasing lead times?	____	____	____	____
33. Is the master schedule stable and not subject to erratic changes?	____	____	____	____
34. Does the master schedule have a frozen zone?	____	____	____	____
35. Does master schedule planning employ rough-cut capacity planning?	____	____	____	____
F. *Customer Order Servicing*				
36. Are customer orders processed promptly from sales to shipment?	____	____	____	____
37. Have objective standards for delivery performances been established?	____	____	____	

	Currently		Desired in the Future	
	Yes	No	Yes	No
38. Are these standards expressed in terms of number of days from receipt to order to shipment?	___	___	___	___
39. Have individual standards been established not only for each product family but also for different market sectors?	___	___	___	___
40. Is there a high ratio of unfilled back orders?	___	___	___	___
41. Is the cost of a back order known?	___	___	___	___
G. *Inventory Accounting*				
42. Are there written inventory objectives as to how much the investment in finished goods, raw materials, and components should be?	___	___	___	___
43. Are there written goals for the desired inventory turnover rates for finished goods, raw materials, and components?	___	___	___	___
44. Are there written policies for production planners, material controllers, and inventory analysts which will guide them in determining the order quantity and the safety stock?	___	___	___	___
45. Are there specific policies for finished goods, raw materials, and components?	___	___	___	___
46. Are there separate policies for groups of items depending on annual usage value (ABC classification)?	___	___	___	___
47. Are materials usually available when needed for the start of production?	___	___	___	___
48. Are inventory records accurate to at least 95%?	___	___	___	___
49. Is immediate inventory status information available?	___	___	___	___

(Continued)

	Currently		Desired in the Future	
	Yes	No	Yes	No
50. Does inventory status information show open orders for both manufactured and purchased items?	___	___	___	___
51. Is there an inventory analysis report which lists excess items in inventory, sequenced by value?	___	___	___	___
52. Is prompt and proper disposition made of materials and parts that have become obsolete as a result of an engineering change?	___	___	___	___
53. Have inventory levels been reduced by profitable disposition of obsolete and excess items?	___	___	___	___
54. Is an ABC analysis performed periodically?	___	___	___	___
55. Is there a documented procedure or system in effect for arriving at make-or-buy decisions?	___	___	___	___
56. Are make-or-buy decisions made on the basis of balancing plant capacity?	___	___	___	___
57. Is a calculation periodically make on the probability of shipping an end item based on the service level of the components for that end item?	___	___	___	___
58. Is safety stock maintained?	___	___	___	___
59. Is an annual physical taken?	___	___	___	___
60. Is there a cycle counting system in place?	___	___	___	___
61. Is inventory physically staged or kitted?	___	___	___	___
62. Is a floating locator system used in the warehouse?	___	___	___	___
63. Is the amount of work in process (WIP) known?	___	___	___	___
64. Is the carry cost rate known?	___	___	___	___
H. *Bill of Material Management*				
65. Are the bills of material complete and accurate?	___	___	___	___

	Currently Yes	Currently No	Desired in the Future Yes	Desired in the Future No
66. Are bills of material available for all products?	___	___	___	___
67. Is there a level-by-level product structure for proper explosion of every bill of material?	___	___	___	___
68. Are all bills of material updated in accordance with the latest engineering revision?	___	___	___	___
69. Are nonsignificant part numbers used?	___	___	___	___
70. Are security controls in place regarding who can change a bill of material?	___	___	___	___
71. Is an indented bill of material available for use?	___	___	___	___
72. Is a periodic evaluation made as to which products can be incorporated into model/option product structures?	___	___	___	___
73. Is an engineering change control system in place with effectivity start–stop dates and run out control?	___	___	___	___
I. *Material Requirements Planning (MRP)*				
74. Is material requirements planning used for items where the demand can be calculated?	___	___	___	___
75. Is the frequency of MRP action notices timed for optimum management control without undue nervousness of the system?	___	___	___	___
76. Are quantities on hand and scheduled receipts correctly reflected in the system?	___	___	___	___
77. Are there action notices for planned order release, open order past due, expedite open order, reschedule open order, and other required actions?	___	___	___	___

(Continued)

	Currently		Desired in the Future	
	Yes	No	Yes	No
78. Are lead times correctly stated and maintained current?				
79. Do material requirements cover the full manufacturing and purchasing planning horizon?				
80. Are there multiple lot-sizing options in the system for application of the optimum technique for each item?				
81. Is MRP run frequent enough to accommodate changes?				
82. Does the MRP system utilize time-phased allocations?				
83. Will the MRP system handle multiple plant locations?				
84. Does the system specify the one of multiple locations at which an allocation has been made?				
J. *Shop Order Release Control*				
85. Are orders released for production so as not to exceed actual shop capacity?				
86. Are backlogs kept off the shop floor?				
87. Are orders scheduled based on latest requirements?				
88. Are orders scheduled to a short cycle (daily or weekly)?				
89. Are only those items scheduled which the factory can make?				
90. Are schedules being followed and due dates being met?				
91. Are pick lists generated with locations at order release time?				
92. Is MRP notified that an order has been released to the shop floor?				
93. Are routings automatically printed by the shop order release function?				

	Currently		Desired in the Future	
	Yes	No	Yes	No
K. *Capacity Requirements Planning*				
94. Are input queue, set-up, run, output queue, and move times available for capacity requirements planning?	___	___	___	___
95. Is available capacity projected and used to produce a dispatch list?	___	___	___	___
96. Are performance (efficiency) factors developed for each work center?	___	___	___	___
97. Are all work centers that have limited capacity included in the capacity requirements planning system?	___	___	___	___
98. Is all work which was not completed during the planned period, replanned in the next or subsequent periods?	___	___	___	___
99. Does the capacity requirements planning system simulate work performance and provide projected completion dates?	___	___	___	___
L. *Purchasing*				
100. Is there an effective system for vendor selection on the basis of quality, delivery, and price?	___	___	___	___
101. Is there a price monitor report which flags price increases for review by purchasing management?	___	___	___	___
102. Does the system track vendor delivery performance (early and late, over and under shipments, quality control rejections)?	___	___	___	___
103. Are blanket orders and releases used to reserve vendor capacity and optimize delivery lead time?	___	___	___	___

(Continued)

	Currently		Desired in the Future	
	Yes	No	Yes	No
104. Are changes in vendor lead times transmitted promptly to the inventory planning/material requirements planning system?	____	____	____	____
105. Is there an effective system for expediting critical items?	____	____	____	____
106. Are quantity discounts and future price increases considered in determining optimum purchase order quantities?	____	____	____	____
107. Are purchase order quantities consistent with inventory policy?	____	____	____	____
M. *Receiving*				
108. Is material forwarded to destination promptly upon receipt?	____	____	____	____
109. Are receiving transactions reported accurately and timely?	____	____	____	____
110. Is there an effective system for receiving material that is not on a purchase order (for example, products returned by customer)?	____	____	____	____
111. Does the system track that material is received, but is pending quality control incoming inspection?	____	____	____	____
112. Are terminals available in receiving to display purchase order data?	____	____	____	____
N. *Shop Floor Control*				
113. Do order release dates precede by the shortest possible interval, the actual date that the first operation will start?	____	____	____	____
114. Are dispatch lists generated daily by work center showing the jobs in priority sequence?	____	____	____	____
115. Is expediting limited to a few problems or rush jobs?	____	____	____	____
116. Are machines idle awaiting work?	____	____	____	____

		Currently		Desired in the Future	
		Yes	No	Yes	No
117.	Is there an overload of work in input queues?	___	___	___	___
118.	Are jobs selected for work so as to minimize set-up time (grouping)?	___	___	___	___
119.	Is feedback from the shop accurate in reporting scrap and production?	___	___	___	___
120.	Are jobs de-expedited as well as expedited?	___	___	___	___
121.	Are shop floor personnel measured (paid) by completion of each operation on a shop order?	___	___	___	___
122.	Is a shortage list used?	___	___	___	___
123.	Does the "wait time" in your manufacturing lead time equal 85% or more?	___	___	___	___
124.	Is shop floor feedback done at least daily?	___	___	___	___
125.	Does a good method exist for handling bulk issues?	___	___	___	___